Introduction to Electromagnetism

Essential Electronics Series

Introduction to Electromagnetism

Martin Sibley
Division of Electronics and Communications
School of Engineering
University of Huddersfield

A member of the Hodder Headline Group
LONDON • SYDNEY • AUCKLAND

First published in Great Britain 1996 by Arnold,
a member of the Hodder Headline Group,
338, Euston Road, London NW1 3BH

© 1996 M Sibley

All rights reserved. No part of this publication may be reproduced or transmitted in any form or by any means, electronically or mechanically, including photocopying, recording or any information storage or retrieval system, without either prior permission in writing from the publisher or a licence permitting restricted copying. In the United Kingdom such licences are issued by the Copyright Licensing Agency: 90 Tottenham Court Road, London W1P 9HE.

British Library Cataloguing in Publication Data
A catalogue record for this book is available from the British Library

ISBN 0 340 64595 4

Typeset in $10\frac{1}{2}/13\frac{1}{2}$ Times by Wearset, Boldon, Tyne and Wear.
Printed and bound in Great Britain by J W Arrowsmith Ltd, Bristol.

To my Grandmother

Mrs E Whittall
1887–1993

She saw many technological changes during her life,
and never failed to explore them.

Series Preface

In recent years there have been many changes in the structure of undergraduate courses in engineering and the process is continuing. With the advent of modularization, semesterization and the move towards student-centred learning as class contact time is reduced, students and teachers alike are having to adjust to new methods of learning and teaching.

Essential Electronics is a series of textbooks intended for use by students on degree and diploma level courses in electrical and electronic engineering and related courses such as manufacturing, mechanical, civil and general engineering. Each text is complete in itself and is complementary to other books in the series.

A feature of these books is the acknowledgement of the new culture outlined above and of the fact that students entering higher education are now, through no fault of their own, less well equipped in mathematics and physics than students of ten or even five years ago. With numerous worked examples throughout, and further problems with answers at the end of each chapter, the texts are ideal for directed and independent learning.

The early books in the series cover topics normally found in the first and second year curricula and assume virtually no previous knowledge, with mathematics being kept to a minimum. Later ones are intended for study at final year level.

The authors are all highly qualified chartered engineers with wide experience in higher education and in industry.

R G Powell
Jan 1995
Nottingham Trent University

Table of contents

Preface		xi
List of symbols		xiii
	Chapter 1 Introduction	1
1.1	Historical background	1
1.2	Atomic structure	3
1.3	Vectors and coordinate systems	4
1.4	Line, surface and volume integrals	5
	Chapter 2 Electrostatic fields	8
2.1	Coulomb's law	8
2.2	Electric flux and electric flux density	10
2.3	The electric field and electric field strength	14
2.4	Electric potential	16
2.5	Equipotential lines	20
2.6	Line charges	23
2.7	Surface charges	29
2.8	Volume charges	35
2.9	Capacitors	36
2.10	Some applications	58
2.11	Summary	61
	Chapter 3 Electromagnetic fields	64
3.1	Some fundamental ideas	64
3.2	Some elementary conventions used in electromagnetism	67
3.3	The Biot–Savart law	69
3.4	Electromagnetic flux, flux density and field strength	71
3.5	Comment	73
3.6	Magnetic field strength and Ampère's circuital law	74
3.7	The force between current-carrying wires – the definition of the ampere	77
3.8	The magnetic field of a circular current element	79
3.9	The solenoid	83
3.10	The toroidal coil, reluctance and magnetic potential	88

3.11	Inductance	91
3.12	Some applications	112
3.13	Summary	115

Chapter 4 Electroconductive fields — 120
4.1	Current flow	120
4.2	Potential and electric field strength	121
4.3	Current density and conductivity	122
4.4	Resistors	124
4.5	Some applications	139
4.6	Summary	141

Chapter 5 Comparison of field equations — 143
5.1	Force fields	143
5.2	Flux, flux density and field strength	144
5.3	Potential and resistance to flux	145
5.4	Energy storage	146
5.5	Force	147
5.6	Summary	149

Chapter 6 Dielectrics — 150
6.1	Electric dipoles and dipole moments	150
6.2	Polarization and relative permittivity	153
6.3	Boundary relationships	157
6.4	Dielectric strength and materials	160

Chapter 7 Ferromagnetic materials and components — 163
7.1	Magnetic dipoles and permanent magnets	163
7.2	Polarization and the B/H curve	164
7.3	Boundary relationships	165
7.4	Iron-cored transformers	167
7.5	Electrical machinery	170
7.6	The magnetic circuit	173

Problems — 177

Index — 183

Preface

This book is concerned with the study of electrostatic, electromagnetic and electroconductive field theory. These three have been unified into electromagnetic field theory or, more simply, electromagnetism. This is fundamental to electrical engineering and, as such, all students of the area should have a proper grounding in the subject.

So, what can we gain from the study of such a fundamental area? Well, there are some phenomena that we cannot explain using circuit theory: capacitive and inductive coupling between conductors; insulation breakdown in transmission lines; and rotating electrical machinery (generators and motors). In addition, electromagnetism is used in microwave ovens; electro-magnetic compatibility; radio and television broadcasting; personal communications; circuit design and high-speed computers, to name but a few! In a book of this length I could not hope to cover all aspects of the subject. Instead, I will present the development of field theory in relation to common electrical circuits and components. In this way, readers should get a perspective on electrical engineering that is different from the traditional circuit theory viewpoint that most of us have.

The book is essentially in two parts. Chapters 2–4 present the basic theory of electrostatic, electromagnetic and electroconductive fields. In these chapters I have applied the theory to different transmission lines. This is a very important point to grasp – it is seldom appreciated that every current-carrying wire generates a magnetic field, and this can couple signals into adjacent conductors. We must also take account of capacitive coupling.

This first part culminates in Chapter 5, which compares the basic relationships that lie behind all three field systems. I hope that this will instil some sense of wonder that these field systems can share the same fundamental relationships.

The second part of the book deals with the effects of dielectrics on capacitors, and ferrous materials on coils. This is done in a strictly fields approach which should give readers an appreciation of the physical effects of electric and magnetic fields.

I wish to acknowledge the support my family and friends have given me throughout the preparation of this book. Thanks also go to the staff and students of the School of Engineering, at the University of Huddersfield; to R

Binns of the University, for taking the photographs of the CRT that appear in Chapters 2 and 3; to the Public Relations Department of the Joint European Torus, for the information in Chapter 3; to the series editor, R G Powell of Nottingham Trent University; and to the publisher, D Ross of Arnold, London.

It is my hope that, after reading this book, readers will have gained an insight into why components and circuitry sometimes fail to behave as designed!

M J N Sibley
Huddersfield, 1995

List of symbols

A	area
$\boldsymbol{B}$	magnetic flux density (vector quantity)
c	speed of light (3×10^8 m s^{-1} in free space)
C	total capacitance
C'	capacitance per unit length
$\boldsymbol{D}$	flux density (vector quantity)
$\boldsymbol{E}$	electric field strength (vector quantity)
f	frequency
$\boldsymbol{F}$	force (vector quantity)
$\boldsymbol{H}$	magnetic field strength (vector quantity)
i	instantaneous current
$i(t)$	time-varying current
I	current
$\boldsymbol{J}$	current density (vector quantity)
L	inductance
L'	inductance per unit length
$\boldsymbol{M}$	magnetic polarization (vector quantity)
N	number of turns on a coil
p	magnetic pole strength
$\boldsymbol{P}$	electrostatic polarization (vector quantity)
q	electronic charge (1.6×10^{-19} C on a single electron/proton)
Q	total charge on a body
$\boldsymbol{r}$	radial unit vector (spherical and cylindrical coordinate sets)
R	resistance
R'	resistance per unit length
S	reluctance
$\tan \delta$	loss tangent
$v(t)$	time-varying voltage
V	potential
V_m	magnetic potential
V_pk	peak value of voltage
X_C	reactance of a capacitor
X_L	reactance of an inductor
$\boldsymbol{x}$	x-direction unit vector (Cartesian coordinate set)
$\boldsymbol{y}$	y-direction unit vector (Cartesian coordinate set)

List of symbols

$\mathbf{z}$	z-direction unit vector (Cartesian coordinate set)
ϵ_0	permittivity of free-space (8.854×10^{-12} F m^{-1})
ϵ_r	relative permittivity
ψ	electric flux
ρ_l	line charge density
ρ_s	surface charge density
ρ_v	volume charge density
ω	angular frequency
μ_0	permeability of free space ($4\pi \times 10^{-7}$ H m^{-1})
μ_r	relative permeability
ϕ	magnetic flux
$\boldsymbol{\phi}$	angular unit vector (spherical and cylindrical coordinate sets)
$\boldsymbol{\theta}$	angular unit vector (spherical coordinate set)
$Ü$	flux linkage
σ	conductivity
χ	electric susceptibility

1 Introduction

This book is concerned with the study of electrostatic, electromagnetic, and electroconductive fields – sometimes referred to as field theory or, more simply, electromagnetism. A knowledge of this subject can help us to explain why a circuit refuses to behave as designed, why components sometimes break down, and what happens in high-frequency circuits. In studying this area, life is made a lot easier if we can think in three dimensions. This is usually a case of drawing adequate diagrams, and practising.

Readers used to circuit theory may wonder why they should study such a discipline. Well, field theory is the study of some of the fundamental laws of Nature. Indeed, electromagnetism was the first theory to unite the sciences of electricity and magnetism. The search is now on to find a Grand Unified Theory that unites all the basic forces of Nature. So, the study of field theory is the study of Nature, and that should interest us.

As we progress with our studies, we will meet some names that have become famous in the field of electrical engineering. Some of these people have had units named after them, and so will be more familiar than others. Before we begin our studies in earnest, let us take a moment to pay our respects to some of the researchers who contributed to electrical engineering as we know it today.

1.1 HISTORICAL BACKGROUND

Electromagnetic field theory is really the result of the union of three distinct sciences. The oldest of these is electrostatics, which was first studied by the Greeks. They discovered that if they rubbed certain substances, they were able to attract lighter bodies to them. One of these substances was amber, whose Greek name is *elektron* – this is where we get the name 'electricity'. It was in 1785 that a French physicist, Charles Augustin de Coulomb (1736–1806), showed that electrically charged materials sometimes attract and sometimes repel each other. This was the first indication that there were two types of charge – positive and negative.

In the late 1700s, two Italians were working on the new science of current electricity. One, Luigi Galvani (1737–1798), was a physiologist and physician who thought that animal tissues generate electricity. Although he was later proved wrong, his experiments stimulated Count Alessandro Volta (1745–1827) to invent the first electric battery in 1800. Most of the early experiments in current electricity were performed on frog's legs – this was a result of Galvani's work.

Later, a favourite party trick was to get a group of people to hold hands, and then connect them to a Voltaic cell (a battery). The cell produced quite a large voltage, which then caused current to flow through the guests. This made them jump uncontrollably! It wasn't until 1833 that the British experimenter Michael Faraday (1791–1867) showed that the current electricity of Volta and Galvani was the same as the electrostatic electricity of Coulomb. Rather than linking these two phenomena, it was shown that current and electrostatic electricity were one and the same thing. (Faraday's contribution is all the more remarkable when it is realized that his theories were formulated by direct experimentation, and not by manipulating mathematics!)

Although the ancient Greeks also knew about magnetism in the form of lodestone, the Chinese invented the magnetic compass and in 1600, William Gilbert of Gloucester laid down some fundamentals. However, it was not until 1785 that Coulomb formulated his law relating the strengths of two magnetic poles to the force between them. Magnetism may have been laid to rest here if it wasn't for the Danish physicist Hans Christian Oersted (1777–1851). It was Oersted who demonstrated to a group of students that a current-carrying wire produces a magnetic field. This was the first sign that electricity and magnetism could be interlinked. This link was strengthened in 1831 by the work of Faraday who showed that a changing magnetic field could induce a current into a wire. It was the French physicist André Marie Ampère who first formulated the idea that the field of a permanent magnet could be due to currents in the material. (We now accept that electrons orbiting the nucleus constitute a current, and this produces the magnetic field.)

We owe our present view of 'field theory' to Faraday who performed many experiments on electricity and magnetism. Although Faraday preferred to work without mathematics, he did introduce the idea of fields in free space. This greatly influenced later workers, and it was in the mid-1800s that the British physicist James Clerk Maxwell (1831–1879) formalized Faraday's results using mathematics. Among other things, Maxwell was able to predict the existence of electromagnetic waves. This work inspired others in the field, such as Oliver Heaviside (1850–1925) who worked on the first transatlantic telegraph cable, as well as predicting the existence of the ionosphere.

The rest, as they say, is history. Due to the work of the German physicist Heinrich Rudolf Hertz (1857–1894) and the Italian engineer Guglielmo Marconi (1874–1937), we are now able to communicate over vast distances. We can also use electrical machinery to make our lives more comfortable. In fact, we owe our current way of life to the hard work of a number of researchers who continually questioned and experimented, carefully recording their results and observations.

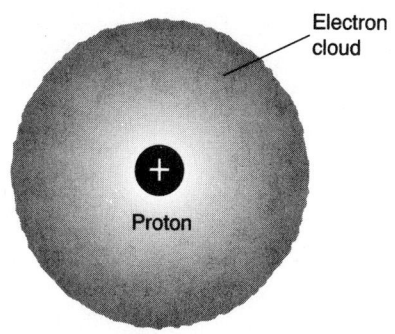

Fig. 1.1 Basic structure of a hydrogen atom

1.2 ATOMIC STRUCTURE

When we learn to drive a car, we do not necessarily need to know exactly how the car works. However, if we do understand how the engine works and why the wheels turn, it can help us to be better drivers. A similar situation occurs with electricity and magnetism – when we use electricity and magnetism we seldom have to worry about exactly how the effects are produced. However, it can make us better engineers if we have an adequate model of what electricity and magnetism are. This is where we have to study the structure of the atom.

Figure 1.1 shows the basic structure of the simplest atom, the hydrogen atom. This atom has one electron that orbits the nucleus containing a single proton. The charge on the electron is equal and opposite to the charge on the proton and has the value

$$q = 1.6 \times 10^{-19} \tag{1.1}$$

with units of coulomb, symbol C.

More complex materials, such as amber for instance, have many atoms held in a crystalline structure. If we rub amber, the friction removes electrons, so leaving the material positively charged. This is the basis of electrostatic electricity. In some materials the electrons are very tightly bound to the nucleus, and considerable energy must be expended to remove an electron. These are insulators.

In a metal atom, the electron in the outermost orbit is not tightly bound to the nucleus. When a number of metal atoms are close to each other, they form a crystalline structure in which these outermost electrons are free to move around; see Fig. 1.2. Now, metals are usually electrically neutral with the number of electrons exactly balancing the number of protons. If we connect a source of electrons to the metal, injected electrons will travel through the lattice. As like charges repel, these electrons force the free electrons away from them. The net effect is to produce a disturbance that travels down the metal. The rate of flow of charge is the electric current. We should note that the mass

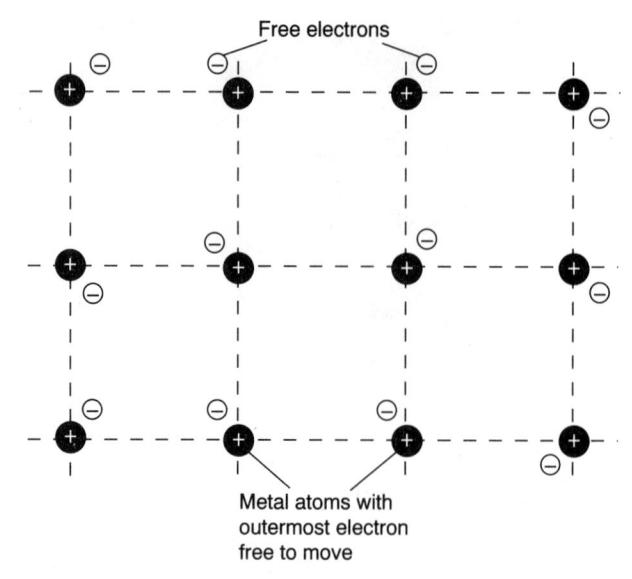

Fig. 1.2 Basic crystalline structure of a metal

of a proton is 1837 times the mass of an electron. Thus conduction in metals is by electron flow.

Let us now turn our attention to magnetism. As we will see in Chapter 3, a current generates a magnetic field. In an atom we can regard the motion of electrons around a nucleus as constituting a current. Thus there will be a magnetic field. In most materials the electron orbit is completely random, and so there is no perceptible magnetic field. However, in some materials, e.g. iron, the electrons can travel in the same general direction. Thus each atom becomes a permanent magnet. When these atoms are part of a crystalline structure, their magnetic fields are randomly distributed, and there are negligible external effects. However, if we subject the material to an external magnetic field, the atomic magnets align themselves to the field. When we remove the field, some of these atomic magnets stay in their new positions, so producing a permanent magnet.

This section has introduced us to some basic ideas about electricity and magnetism. Although this discussion has been very simplistic in form, the models we developed will be useful in later chapters. We will now consider vector notation, and briefly examine some coordinate systems.

1.3 VECTORS AND COORDINATE SYSTEMS

When we use a thermometer, we read the temperature off a graduated scale. The temperature of a body is independent of direction (it is simply measured at a certain point) and so it is known as a scalar quantity. Scalar quantities are those that have no direction associated with them.

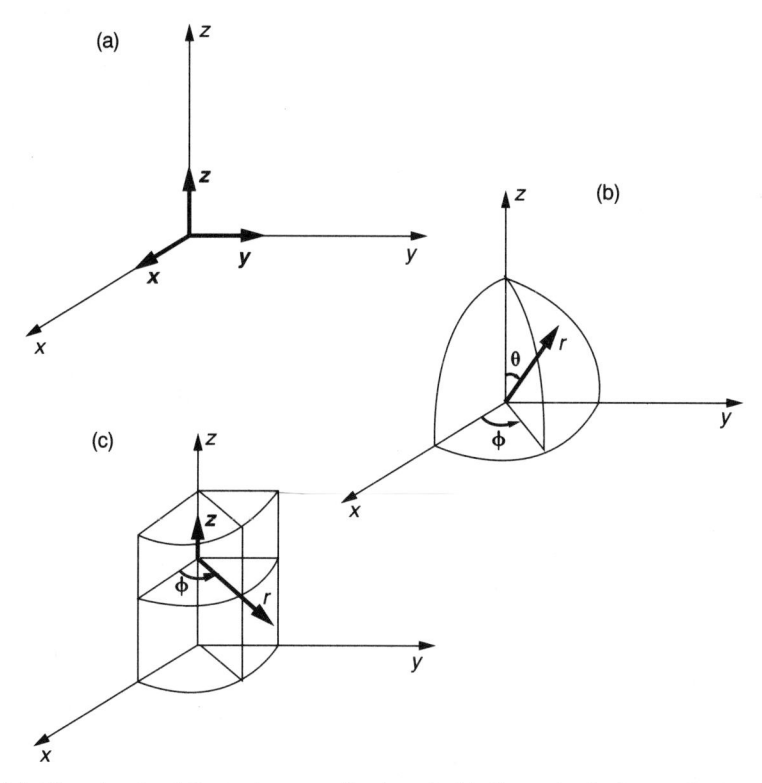

Fig. 1.3 (a) The standard Cartesian coordinate set; (b) the spherical coordinate set; and (c) the cylindrical coordinate set

If we push an object, we have to exert a force on it. This force has direction associated with it – we could push the object to the left, to the right, or in any direction we choose. The force is a vector quantity because it has magnitude and direction.

At this point, we could launch into a discussion of vector theory – addition, multiplication, etc. Unfortunately this would complicate matters, and mask the underlying ideas. Instead we will avoid vector algebra in favour of discussion and reasoning. In spite of this, Fig. 1.3 shows the standard Cartesian, spherical, and cylindrical systems that we will use as we progress with our studies. (We will use unit vectors in most of the text, however. This is to help readers get used to vector notation, which will aid future studies.)

1.4 LINE, SURFACE AND VOLUME INTEGRALS

This book assumes that readers are familiar with basic integration and differentiation. However, we will come across many instances where we need the line, surface and volume integrals. As most readers will not be very familiar with these integrals, this section deals with their definition and application.

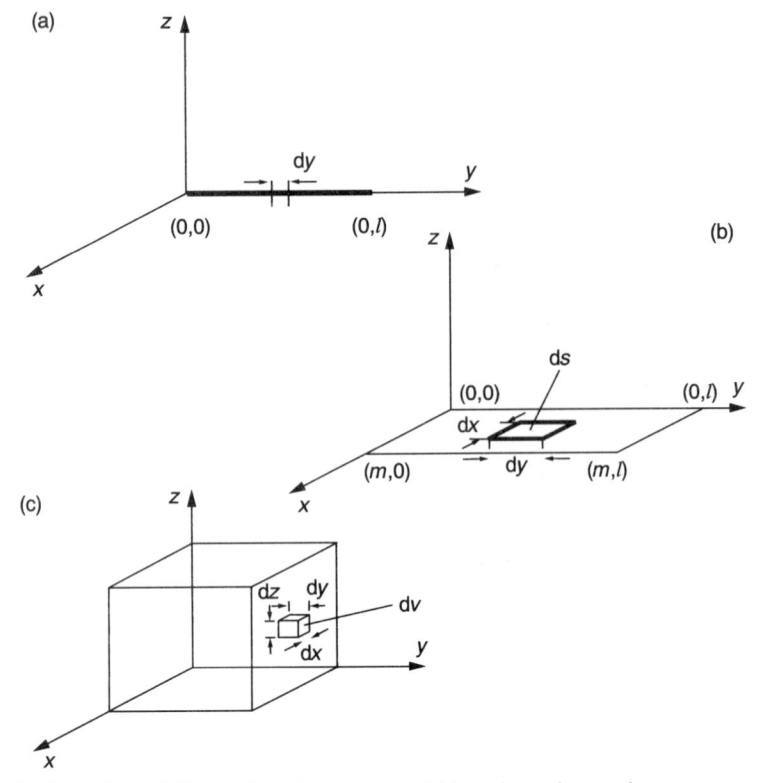

Fig. 1.4 (a) Line integral; (b) surface integral; and (c) volume integral

Figure 1.4(a) shows a line of length l. Let us consider a small incremental section of the line, of length dy. Now, the left-hand end of the line is at the origin of a Cartesian coordinate set, and the line lies along the y-axis. When we do the line integral, we effectively add together the lengths of the incremental section as we move it along the line. This is represented by

$$\text{length} = \int_0^l dy$$
$$= |y|_{y=0}^{y=l}$$
$$= l$$

So, the line integral merely gives us the length of the line. As regards the surface integral, Fig. 1.4(b) shows a square lying in the xy-plane. Let us take a small incremental section of area ds. This area is given by

$$ds = dx\, dy$$

Now, when we perform a surface integral, we effectively move ds across the whole of the square. We can split this into two parts: we can integrate with respect to y to give a line of thickness dx and length l, and then integrate with respect to x to give the total area. So,

$$\text{area} = \int_0^m \left(\int_0^l dy \right) dx$$
$$= \int_0^m l\, dx$$
$$= ml$$

The volume integral follows a similar procedure: we consider a small incremental volume, and integrate with respect to x, y and z. This is shown in Fig. 1.4(c). So, the volume integral is

$$\text{volume} = \int_0^n \left(\int_0^m \left\{ \int_0^l dy \right\} dx \right) dz$$
$$= \int_0^n \left(\int_0^m l\, dx \right) dz$$
$$= \int_0^n lm\, dz$$
$$= lmn$$

Although we have confined ourselves to a Cartesian coordinate set, we could have considered the spherical or cylindrical sets. With these sets the same basic principle applies – consider a small incremental section, and then integrate with respect to the relevant coordinates. Problems 1.1–1.5 should provide readers with practice!

2 Electrostatic fields

Most of us are familiar with the phenomenon of electrostatic discharge: lightning strikes; sparks from nylon clothing; and sparks from nylon carpet. It may be thought that the study of static electric fields has little to offer the electrical engineer. After all, we are taught that electrons flow in conducting materials, and so why should we concern ourselves with the study of static charges? However, as we shall see later in this chapter, electrostatics introduces several ideas that will be very helpful when we consider capacitors and, ultimately, transmission lines.

2.1 COULOMB'S LAW

As we have seen in Chapter 1, electronic charge comes in two forms: negative charge from an electron, and positive charge from a proton. In both cases, a single isolated charge has a charge of 1.6×10^{-19} coulomb. If there are two charges close to each other, they tend to repel each other if the charges are alike, or attract each other if they are dissimilar. Thus we can say that these charges exert a force on each other.

Charles Augustin de Coulomb (1736–1806) determined by direct experimental observation that the force between two charges is proportional to the product of the two charges, and inversely proportional to the square of the distance between them. In terms of the SI units, the force between two charges, a vector quantity, is given by

$$F = \frac{q_1 q_2}{4\pi \epsilon r^2} r \qquad (2.1)$$

where F is the force between the charges (N)
 q_1 and q_2 are the magnitudes of the two charges (C)
 ϵ is a material constant (F m^{-1})
 r is the distance between the two charges (m)
and r is a unit vector acting in the direction of the line joining the two charges – the radial unit vector.

This is Coulomb's law. The force, as given by Equation (2.1), is positive (i.e.

Electron cloud   Proton

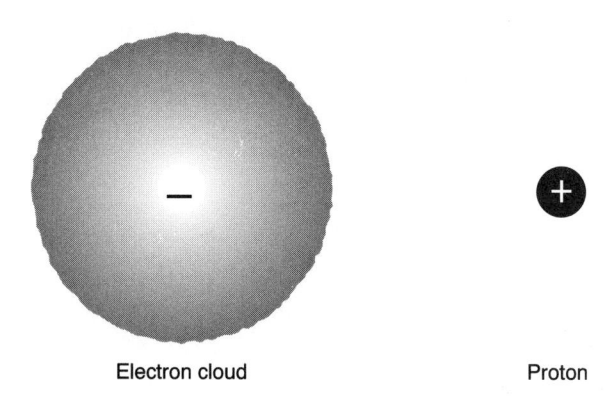

Fig. 2.1 Two separate charges in free space

repulsive) if the charges are alike, and negative (i.e. attractive) if the charges are dissimilar (see Fig. 2.1). As Equation (2.1) shows, the force between the charges is inversely dependent on a material constant, ϵ, the permittivity. Good insulators have very high values of permittivity, typically 10 times that of air for glass, and so the electrostatic force is correspondingly smaller.

If no material separates the charges, i.e. if they are in a vacuum, the permittivity has the lowest possible value of $8.854 \times 10^{-12}\,\mathrm{F\,m^{-1}}$ or $1/36\pi \times 10^{-9}\,\mathrm{F\,m^{-1}}$. (These rather obscure values result from the adoption of the SI units.) As permittivity has such a low value, it is more usual to normalize the permittivity of a material to that of free space. This normalized permittivity is commonly known as the relative permittivity, ϵ_r, given by

$$\epsilon_r = \frac{\epsilon}{\epsilon_0} \qquad (2.2)$$

With this form of normalization, ϵ_r ranges from 1.0006 for air, to 5–10 for a good insulator such as glass.

Before we consider an example, it is worth examining Coulomb's law in greater detail. One of the first things we should note is that Coulomb's law incorporates the inverse square law, i.e. the force is inversely proportional to the square of the distance between the charges. This relationship is more commonly found when considering gravitational fields; however, we will meet it again when we study magnetic fields. Another point worth noting is the presence of $4\pi r^2$ in the denominator of Equation (2.1). This is simply the surface area of a sphere, and we will see why this is so when we consider electric flux in the next section.

Example 2.1

Determine the force between two identical charges, of magnitude 10 pC, separated by a distance of 1 mm situated in free space. What is the force if the separation is reduced to 1 µm?

Solution

The force between two charges is given by Coulomb's law, Equation (2.1), as

$$F = \frac{q_1 q_2}{4\pi\epsilon r^2} \mathbf{r}$$

Thus,

$$F = \frac{10 \times 10^{-12} \times 10 \times 10^{-12}}{4 \times \pi \times 8.854 \times 10^{-12} \times (1 \times 10^{-3})^2} \mathbf{r}$$

$$= 0.9 \times 10^{-6} \mathbf{r} \text{ N} \quad \text{(repulsive)}$$

If we reduce the separation to 1 μm, the force increases to 0.9 N.

By way of contrast, if we consider a hydrogen atom, the single electron is in orbit around the single proton at a minimum distance of 5.3×10^{-11} m. This gives an attractive force of 8.2×10^{-8} N. Thus we can see that the electrostatic force in an atom is very small.

In this section we have seen that charges exert a force on each other. This force is repulsive if the charges are alike, and attractive if the charges are unlike. This effect raises the question: how does one charge 'know' that the other is present? To answer this we will introduce the idea of electric flux.

2.2 ELECTRIC FLUX AND ELECTRIC FLUX DENSITY

One definition of flux is that it is the flow of material from one place to another. Some familiar examples of flow are water flowing out of a tap or spring; air flow from areas of high pressure to low pressure; and audio waves flowing outward from a source of disturbance. In general, we can say that flux flows away from a source, and towards a sink.

If we adapt this to electrostatics, we can say that a positive charge is a source of electric flux, and a negative charge acts as a sink. We must exercise extreme caution here. Nothing physically flows out of positive charges – a charge does not run out of electric flux! What we are doing is adapting the general definition of flux, so that we can visualize what is happening. If we consider isolated point charges, we can draw a diagram as in Fig. 2.2. (A point charge is simply a physically small charge, or collection of charges. This raises the question of how small is small? The answer lies with relative sizes. Relative to the distance between the Earth and the Sun, the height of Mount Everest is insignificant. Similarly, we can regard a collection of individual charges, arranged in a 10 nm diameter sphere, as a point charge when viewed from 10 metres away.)

Now, what happens to the distribution of electric flux if we bring two positive charges together? As the charges are both sources of electric flux, the fluxes

2.2 Electric flux and electric flux density

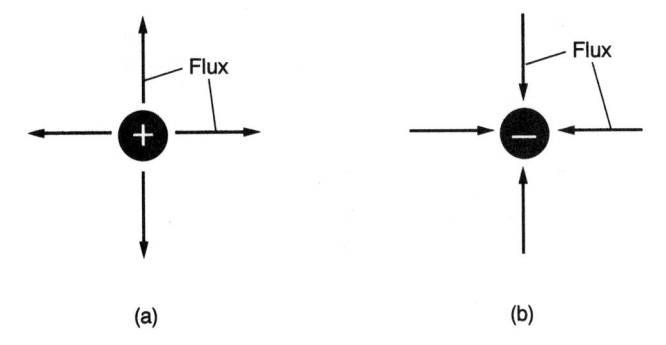

Fig. 2.2 Flux radiation from isolated point charges: (a) a positive charge; and (b) a negative charge

repel each other to produce the distribution shown in Fig. 2.3. One of the main things to note from this diagram is the distortion of the lines of flux in the space between the charges. This causes the force of repulsion between the two charges, in agreement with Coulomb's law.

If we now return to Coulomb's law, we can rewrite it as

$$\mathbf{F} = \frac{q_1}{4\pi r^2} \frac{1}{\epsilon} q_2 \mathbf{r} \tag{2.3}$$

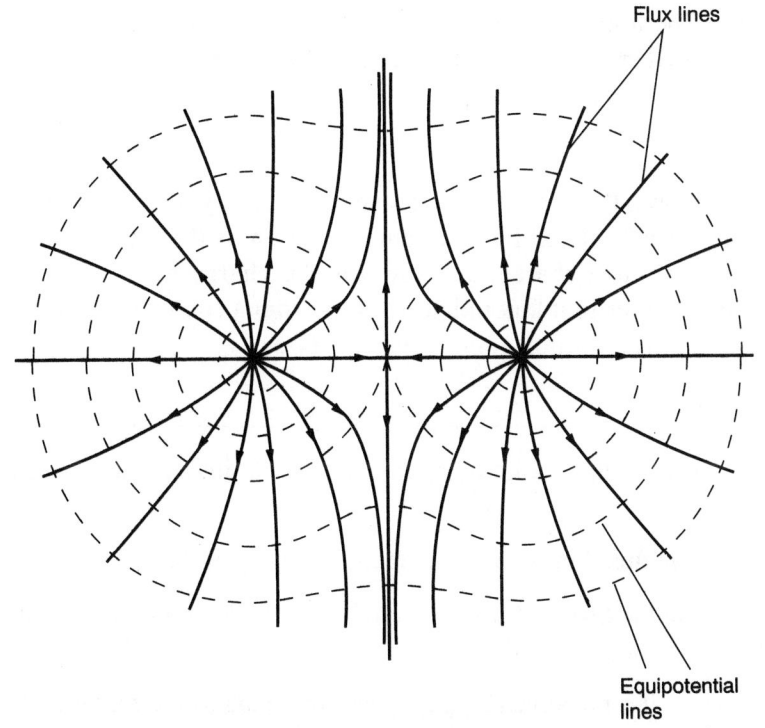

Fig. 2.3 Distribution of flux due to two positive point charges in close proximity

12 Electrostatic fields

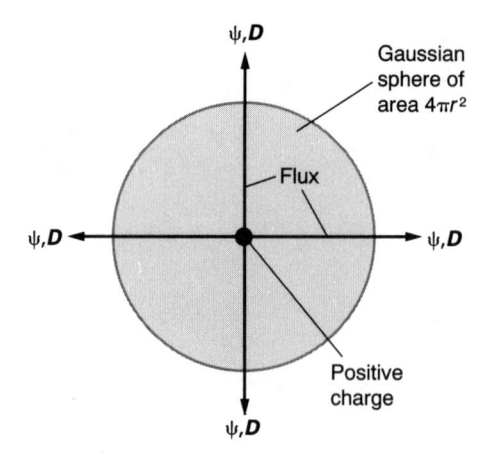

Fig. 2.4 Relating to the definition of flux density

The first term in Equation (2.3) consists of the electronic charge, q_1, divided by the surface area of a sphere, $4\pi r^2$. Thus $q_1/4\pi r^2$ has units of $C\,m^{-2}$, and would appear to be a surface density of some sort – the flux density. To explain this, we must use Gauss' law (Karl Friedrich Gauss, 1777–1855) which states that the flux through any closed surface is equal to the charge enclosed by that surface.

Figure 2.4 shows an imaginary spherical surface surrounding an isolated point charge. Application of Gauss' law shows that the flux, ψ, radiating outwards in all directions has a value of q_1 – the amount of charge enclosed by the sphere. The area of the Gaussian surface is simply that of a sphere, i.e. a surface area of $4\pi r^2$. Thus we get a flux density, $\boldsymbol{D}$, of

$$\boldsymbol{D} = \frac{q_1}{4\pi r^2}\boldsymbol{r} \tag{2.4}$$

As Fig. 2.4 shows, the flux density is a vector quantity, i.e. it has direction and magnitude. Specifically, $\boldsymbol{D}$ has the same direction as the flux – **away** from the charge for a positive charge, and **towards** the charge for a negative charge. Convention dictates that flux radiating away from a charge is positive, whereas the opposite is true for flux going towards a body.

Example 2.2

(1) Determine the flux radiating from a positive point charge of magnitude 100 pC.

(2) What is the flux density at a distance 10 mm from the charge?

(3) Determine the flux that flows through an area 200 mm^2 on the surface of a 1 m radius Gaussian sphere.

(4) Repeat (3) if a negative charge of the same magnitude replaces the positive charge.

Solution

1. Application of Gauss' law shows that the flux from the 100 pC charge is simply the magnitude of the charge. Thus,

    $\psi = 100 \text{ pC}$

2. At a radius of 10 mm, the total flux is still 100 pC. However, the surface area of the sphere is $4\pi r^2$ where r is the radius of the sphere. So, as the radius of the Gaussian sphere is 10 mm, we get a flux density of

    $$D = \frac{100 \times 10^{-12}}{4\pi(10 \times 10^{-3})^2} r$$

    $= 7.96 \times 10^{-8} r \quad C\,m^{-2}$ in a radial direction

3. We now need to find the flux through a 200 mm² area on the surface of a 1 m radius Gaussian sphere. So, the flux density at this radius is

    $$D = \frac{100 \times 10^{-12}}{4\pi 1^2} r$$

    $= 7.96 \times 10^{-12} r \quad C\,m^{-2}$

    As this flux density flows through the 200 mm² surface, the flux is

    $\psi = 7.96 \times 10^{-12} \times 200 \times 10^{-6}$
    $= 1.59 \times 10^{-15} \text{ C}$

4. We now replace the positive charge by a negative one of the same magnitude. As the charge is numerically the same, the magnitudes of all the quantities will be the same. However, as the charge is negative, we have to put a minus sign in front of the answers. So, the flux from the charge is

    $\psi = -100 \text{ pC}$

    the flux density at 10 mm is

    $D = -7.96 \times 10^{-8} r \quad C\,m^{-2}$

    and the flux through the 200 mm² surface is

    $\psi = -1.59 \times 10^{-15} \text{ C}$

This section has shown us that positive electric charges radiate flux, whereas negative charges attract flux to them. This model enables us to draw field plots such as that in Fig. 2.3. Although such plots can help us in visualizing the field surrounding a point charge, we are usually more concerned with the force on a charge due to the presence of a fixed charge. This is where the idea of an electric field becomes useful.

2.3 THE ELECTRIC FIELD AND ELECTRIC FIELD STRENGTH

As we saw in the previous section, we can write Coulomb's law as

$$F = \frac{q_1}{4\pi r^2} \frac{1}{\epsilon} q_2 r$$

If we use the definition of electric flux density given by Equation (2.4) we can write

$$F = \frac{D}{\epsilon} q_2 \tag{2.5}$$

In Equation (2.5) we can see that the force on the charge q_2 is directly proportional to the factor D/ϵ. This factor has units of newton per coulomb, i.e. $N\,C^{-1}$. Thus we can regard this quantity as the force on a unit of charge. To emphasize this more, we introduce a new parameter known as the electric field strength, E, defined as

$$E = \frac{D}{\epsilon} \tag{2.6}$$

We can now write Coulomb's law as

$$F = q_2 E$$

or, more generally,

$$F = q E \tag{2.7}$$

From this equation we can see that the force is directly dependent on the electric field strength, also known as the electric field intensity. We should note that, as E is directly proportional to D, it is also a vector quantity.

The field strength is an important parameter in that it introduces us to the idea of a force field. Figure 2.5(a) shows an isolated point charge at the centre

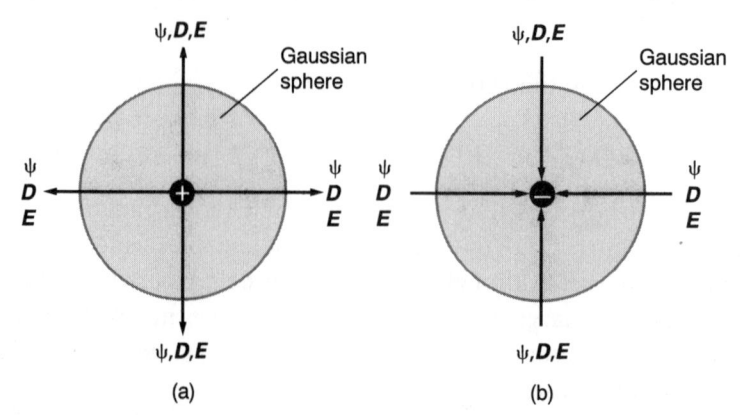

Fig. 2.5 Flux, flux density and electric field strength in (a) a positive charge field; and (b) a negative charge field

of a Gaussian sphere. As this figure shows, electric flux, ψ, radiates outwards from the charge. Also shown are the electric flux density and electric field strength vectors.

Let us now introduce a positive test charge of 1 C. This charge will experience a repulsive force acting in a radial direction – the direction of the **E** field. As this test charge is 1 C, the magnitude of the force will also be the magnitude of the **E** field. Thus the lines of flux are also the lines of force emanating from the charge. A similar situation arises with a negative charge: Fig. 2.5(b). So, we can say that a force field surrounds each charge, and that the field is repulsive if the charges are alike, and attractive if the charges are dissimilar.

Example 2.3

Determine the flux density and electric field strength at a distance of 0.5 m from an isolated point charge of +10 μC. If an identical charge is placed at this point, determine the force it experiences. Assume that the charge is in air.

Solution

Let us place the point charge at the centre of a Gaussian sphere of radius 0.5 m. Now, from Gauss' law the total flux through the sphere is equal to the enclosed charge, i.e.

ψ = 10 μC

The area of the Gaussian sphere is $4\pi r^2$, and so the flux density at 0.5 m is

$$\mathbf{D} = \frac{10 \times 10^{-6}}{4\pi 0.5^2}\mathbf{r}$$

$$= 3.2 \times 10^{-6}\,\mathbf{r}\quad \mathrm{C\,m^{-2}}$$

The strength of the electric field at this radius is

$$\mathbf{E} = \frac{\mathbf{D}}{\epsilon_0}\mathbf{r}$$

$$= \frac{3.2 \times 10^{-6}}{8.854 \times 10^{-12}}\mathbf{r}$$

$$= 3.6 \times 10^5\,\mathbf{r}\quad \mathrm{N\,C^{-1}}$$

Now, if we introduce a 10 μC point charge at this distance, the charge will experience a repulsive force of

$$\mathbf{F} = q\mathbf{E}$$
$$= 10 \times 10^{-6} \times 3.6 \times 10^6\,\mathbf{r}$$
$$= 3.6\,\mathbf{r}\quad \mathrm{N}$$

As a matter of interest, if we halve the distance, we get

$$D = \frac{10 \times 10^{-6}}{4\pi 0.25^2} r$$
$$= 12.7 \times 10^{-6} r \quad C\,m^{-2}$$
$$E = \frac{12.7 \times 10^{-6}}{8.854 \times 10^{-12}} r$$
$$= 1.44 \times 10^6 r \quad N\,C^{-1}$$

and,

$$F = 1.44 \times 10^6 \times 10 \times 10^{-6} r$$
$$= 14.4\, r \quad N$$

This example has shown that, in spite of the small values of charge, and the large distance between them, the electrostatic force can be quite high.

We can now write three, equivalent, forms of Coulomb's law:

$$F = \frac{q_1 q_2}{4\pi\epsilon r^2} r \quad N$$

$$F = \frac{D}{\epsilon} q_2 \quad N$$

and,

$$F = q_2 E \quad N$$

where electric flux density is

$$D = \frac{q_1}{4\pi r^2} r \quad C\,m^{-2}$$

and electric field strength is

$$E = \frac{D}{\epsilon} = \frac{q_1}{4\pi\epsilon r^2} r \quad N\,C^{-1}$$

Let us now turn our attention to electric potential, a term that is usually associated with circuit theory.

2.4 ELECTRIC POTENTIAL

We often come across the term potential when applied to the potential energy of a body, or the potential difference between two points in a circuit. In the former case, the potential energy of a body is related to its height above a certain reference level. Thus a body gains potential energy when we raise it to a

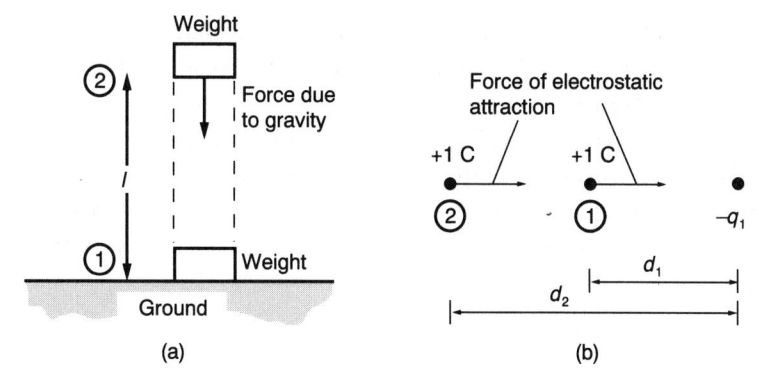

Fig. 2.6 (a) Potential energy in a gravitational field; and (b) potential energy in an electrostatic field

higher level. This gain in energy is equal to the work done against an attractive force, gravity in this example. Figure 2.6(a) shows this situation.

As Fig. 2.6(a) shows, the body is placed in an attractive, gravitational force field. So, if we raise the body through a certain distance, we have to do work against the gravitational field. The difference in potential energy between positions 1 and 2 is equal to the work done in moving the body from 1 to 2, a distance of l metres. This work done is given by,

$$F \times l = m \times 9.81 \times l$$

where m is the mass of the body (kg), and 9.81 is the acceleration due to gravity (m s^{-2}). (Although the effects of gravity vary according to the inverse square law, the difference in gravitational force between positions 1 and 2 is small. This is because the Earth is so large. Thus we can take the gravitational field to be linear in form, and so this equation holds true.)

In an electrostatic field, we have an electrostatic force field instead of a gravitational force field. However, the idea of potential energy is the same. Let us consider the situation in Fig. 2.6(b). We have a positive test charge of 1 C, at a distance d_1 from the fixed negative charge, $-q_1$. This test charge will experience an attractive force whose magnitude we can find from Coulomb's law. Now, if we move the test charge from position 1 to position 2, we have to do work against the field. If the distance between positions 1 and 2 is reasonably large, the strength of the force field decreases as we move away from the fixed charge. Thus we say that we have a non-linear field.

Now, as the field decreases when we move away from the fixed charge, let us move the test charge a very small distance, dr. The electric field strength will hardly alter as we move along this small distance. So, the work done against the field in moving the test charge a small distance dr will be given by

$$\begin{aligned}
\text{work done} &= \text{force} \times \text{distance} \\
&= -F \times dr \\
&= -1 \times E \times dr
\end{aligned} \tag{2.8}$$

18 Electrostatic fields

(The presence of the negative sign is due to the fact that we are moving away from the charge, whereas the electrostatic force acts towards the charge, i.e. in the opposite direction.)

We can move from position 1 to position 2 in very tiny steps so that the E field hardly varies with each step. With each step we take, we will do a small amount of work against the field. To find the total amount of work done, and hence the potential difference, we can integrate Equation (2.8) with respect to r, with d_1 and d_2 as the limits. Thus,

$$\text{total work done} = -\int_{d_1}^{d_2} E \, dr$$

$$= -\int_{d_1}^{d_2} \frac{-q_1}{4\pi\epsilon r^2} \, dr$$

$$= +\frac{q_1}{4\pi\epsilon} \int_{d_1}^{d_2} \frac{1}{r^2} \, dr$$

$$= +\frac{q_1}{4\pi\epsilon} \left| -\frac{1}{r} \right|_{d_1}^{d_2}$$

$$= +\frac{q_1}{4\pi\epsilon} \left\{ \frac{1}{d_1} - \frac{1}{d_2} \right\}$$

In electrical engineering, we use the symbol V for potential. Thus the potential difference between positions 1 and 2 is

$$V_{12} = \frac{q_1}{4\pi\epsilon} \left\{ \frac{1}{d_1} - \frac{1}{d_2} \right\} \tag{2.9}$$

(Although the units of work are joules, we tend to use the volt (named after Count Alessandro Volta, 1745–1827, the Italian physicist who invented the first electric battery in 1800) as the unit of potential.)

Let us take a moment to examine Equation (2.9) more closely. In particular, let us look at the term in the brackets. The major question is whether this term is positive or negative. If d_1 and d_2 are of the same order of magnitude, we might find it difficult to decide. However, if we make d_1 very, very small, and d_2 very, very large, the term in the brackets should be positive. As a check, let us take $d_1 \to 0$ and $d_2 \to \infty$. Thus,

$$V_{12} = \frac{q_1}{4\pi\epsilon} \left\{ \frac{1}{0} - \frac{1}{\infty} \right\}$$

$$= \frac{q_1}{4\pi\epsilon} (\infty - 0)$$

$$= \frac{q_1}{4\pi\epsilon} \infty$$

2.4 Electric potential

Although this quantity is clearly very big, it is also very definitely positive. This confirms that we have to do work against the field in moving the test charge away from the negative charge q_1.

Before we consider an example, let us return to Equation (2.9) again. We can recast this equation as

$$V_1 - V_2 = \frac{q_1}{4\pi\epsilon}\frac{1}{d_1} - \frac{q_1}{4\pi\epsilon}\frac{1}{d_2}$$

from which we can infer that

$$V_1 = \frac{q_1}{4\pi\epsilon}\frac{1}{d_1} \tag{2.10}$$

and,

$$V_2 = \frac{q_1}{4\pi\epsilon}\frac{1}{d_2} \tag{2.11}$$

These voltages are the absolute potentials at points 1 and 2 respectively. We can also find these potentials by moving the test charge from infinity to positions 1 or 2. We must take great care with the minus sign when calculating the absolute potential.

Example 2.4

Determine the absolute potential at a distance 0.2 m from an isolated point charge of 10 µC. Hence determine the potential difference between this point and another at 10 m from the charge.

Solution

The absolute potential is defined as the work done against the field in moving a positive 1 C test charge from infinity to a point in the field. So, the small amount of work done, dV, in moving distance dr is

$$\begin{aligned}dV &= -\text{force} \times dr \\ &= -1 \times E \times dr \\ &= -E\,dr\end{aligned}$$

(Note that this gives

$$E = -\frac{dV}{dr}$$

and so E can have alternative units of V m^{-1}. We will return to this very important point later.)

Thus the total work done in moving the charge from infinity to 0.2 m from the fixed charge, the potential, is

20 Electrostatic fields

$$\int_0^V dV = -\int_\infty^{0.2} \frac{10 \times 10^{-6}}{4\pi\epsilon_0 r^2} dr$$

Therefore,

$$V = -\frac{10 \times 10^{-6}}{4\pi\epsilon_0} \left| -\frac{1}{r} \right|_\infty^{0.2}$$

$$= \frac{10 \times 10^{-6}}{4\pi\epsilon_0} \frac{1}{0.2}$$

$$= 4.5 \times 10^5 \text{ volt} \quad \text{at } 0.2 \text{ m}$$

By following a similar procedure, the potential at 10 m from the charge is

$$V = 9 \times 10^3 \text{ volt}$$

Thus the potential difference between 0.2 m and 10 m is

$$V_{12} = 4.41 \times 10^5 \text{ volt}$$

Before we continue, it is worth stressing again that the units of the E field are also V m^{-1}. We will use these units when we come to consider capacitors in Section 2.9.

2.5 EQUIPOTENTIAL LINES

Let us consider the three paths A, B, and C shown in Fig. 2.7(a). All of these paths link the points 1 and 2, but only path A does so directly. Now, let us take the circular lines in Fig. 2.7(a) as the contours on a hill. In moving from position 1 to position 2 by way of path A, we clearly do work against gravity. The work done is equal to the gain in potential energy which, in turn, is equal to the gravitational force times the change in vertical height. (This is shown in Fig. 2.7(b).)

Now let us take path B. We initially walk left from position 1, around the contour line, to a point directly below position 2. As we have moved around a contour line, we have not gained any height, and so the potential energy remains the same, i.e. we have not done any work against gravity. We now have to walk uphill to position 2. In doing so we do work against gravity equal to the gain in potential energy. This gain in potential energy is clearly the same as with path A. (Although we have to do more physical work in travelling along path B, the change in potential energy is the same.) If we use path C, the same argument holds true. So, we can say that the work done against gravity is independent of the path we take.

Let us now turn our attention to the electrostatic field in Fig. 2.8. As with the contour map, we have three different paths. As we have just seen, we do no work against the field when we move in a circular direction. We only do work

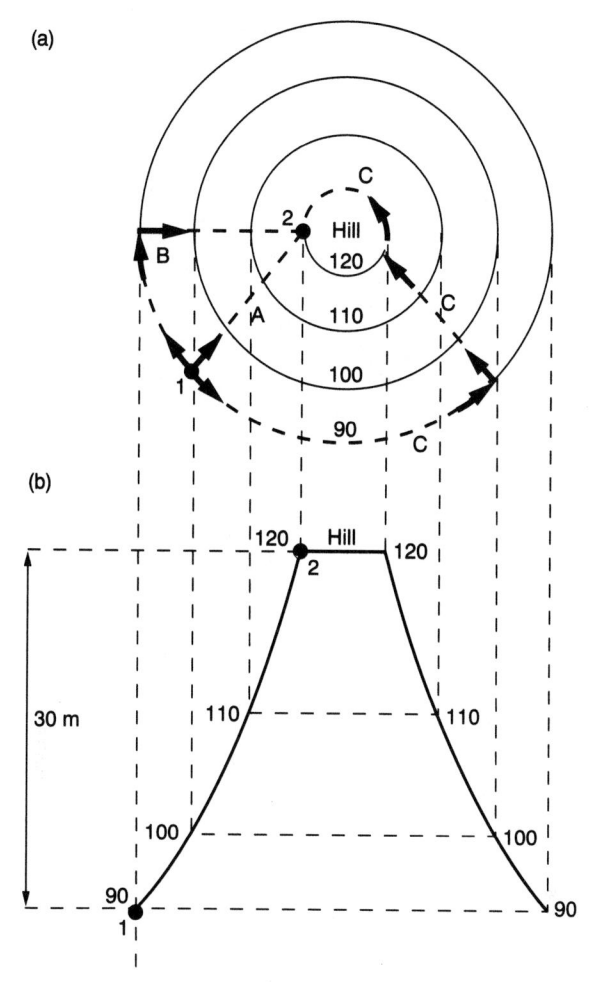

Fig. 2.7 (a) Contour map for a circular hill; and (b) side view of hill

when we move in a radial direction. Thus the potential difference between points 1 and 2 is independent of the exact path we take. This implies that we do no work against the field when we move around the plot in a circular direction. Thus the circular 'contours' in Fig. 2.8 are lines of equal potential, or equipotential lines.

We should be careful when using the term equipotential lines. This is because we are considering a point charge, and so the equipotential surfaces are actually spheres with the charge at their centre. As we are not yet able to draw in a three-dimensional holographic world, we have to make do with two-dimensional diagrams drawn on pieces of paper!

Example 2.5

An isolated point charge, of 20 pC, is situated in air. Plot the 1, 2, and 3 V equipotentials.

22 Electrostatic fields

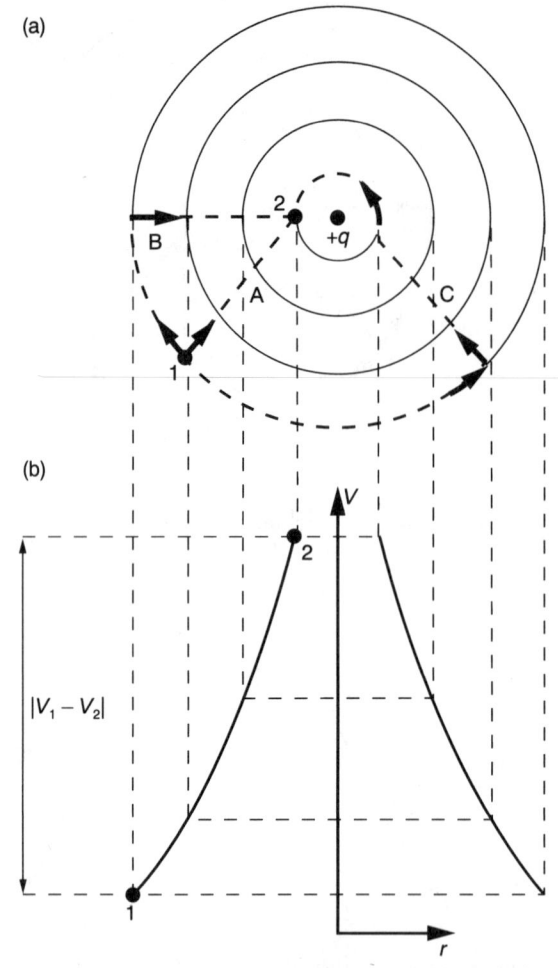

Fig. 2.8 (a) 'Contour' map for a positive point charge; and (b) variation in potential as a function of distance from charge

Solution

To plot the equipotentials, we need to find the radius of the 1, 2, and 3 V potential spheres. Now, the absolute potential of a point in an E field is

$$V = \frac{q}{4\pi\epsilon_0} \frac{1}{r}$$

and so,

$$r = \frac{q}{4\pi\epsilon_0 V}$$

Thus,

$r_{1\,V} = 18$ cm
$r_{2\,V} = 9$ cm

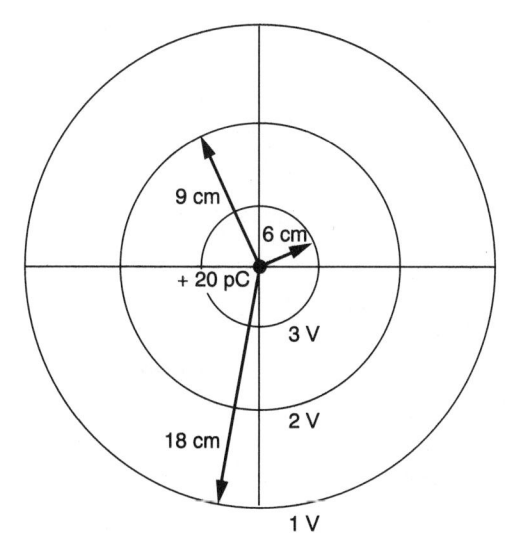

Fig. 2.9 1, 2, and 3 volt equipotentials surrounding a +20 pC point charge

and,

$r_{3V} = 6$ cm

These lines are plotted in Fig. 2.9.

2.6 LINE CHARGES

So far we have only considered point charges. This is very useful when introducing the fundamental laws we have considered. However, we rarely meet point charges in reality. Instead we come across lines of charge, charged surfaces, and charged objects. Thus we have to deal with charge distributions in one, two, and three dimensions. This is where things tend to get a little complicated, as we have to think in three dimensions. In such cases it is essential to draw diagrams that help us visualize the situation.

Let us consider a long piece of wire that is charged by some means. Now, electric flux will radiate outwards from this line of charge. The direction of this flux will be away from the line in a radial direction. If we only consider the central part of the wire, we can ignore what happens at the end of the line, and so the flux distribution is as shown in Fig. 2.10.

If we apply Gauss' law, we can say that the total flux emanating from the wire is equal to the charge enclosed by an imaginary Gaussian surface. In this case, the Gaussian surface will be an open-ended tube with the wire placed along the central axis of the tube (Fig. 2.10). To find the flux density, and hence the electric field strength, we can use Gauss' law, or we can use a more rigorous mathematical approach. Both techniques are presented here.

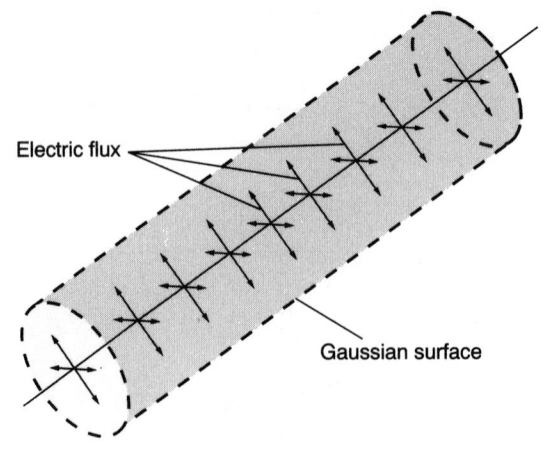

Fig. 2.10 Radiation of electric flux from a line of charge

Gauss' law approach

Let us consider the line charge and Gaussian surface as shown in Fig. 2.10. The charge is distributed along the length of the wire, and so let us introduce a line charge density given by the total charge, Q, divided by the length of the line, L, i.e.

$$\rho_l = \frac{Q}{L} \tag{2.12}$$

If we consider a unit length of wire (1 metre) we get a total flux of

$$\psi = \rho_l \times 1 \quad \text{coulomb} \tag{2.13}$$

Now, the flux density is the flux divided by the surface area of the Gaussian surface. As the Gaussian surface is a tube, the surface area is the circumference of the tube times the length, i.e.

area = $2\pi r \times l$

Thus the flux density is

$$\boldsymbol{D} = \frac{\rho_l}{2\pi r}\boldsymbol{r} \tag{2.14}$$

and the electric field strength is

$$\boldsymbol{E} = \frac{\rho_l}{2\pi \epsilon r}\boldsymbol{r} \tag{2.15}$$

The equipotential surfaces will be coaxial tubes that have the wire along the centre line of the tubes (Fig. 2.11(a)). So, if we move in a direction parallel to the wire, we do no work against the field, indicating that we can ignore travel

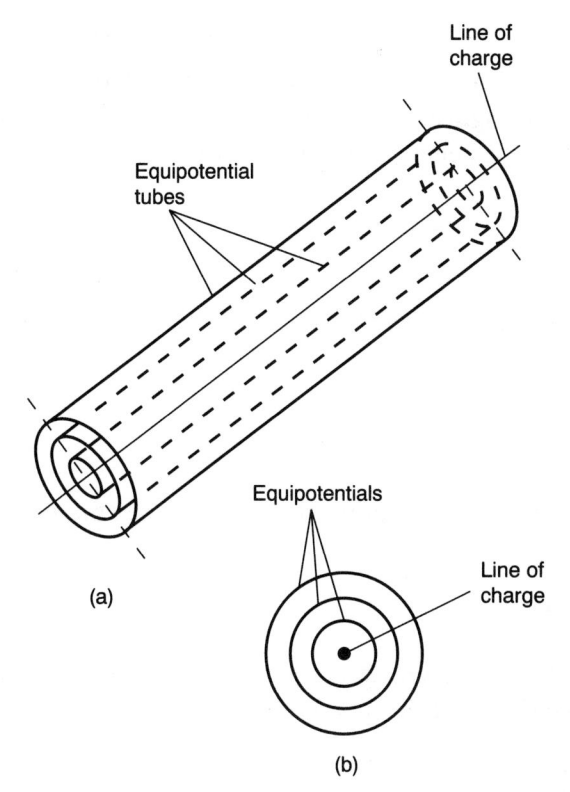

Fig. 2.11 (a) Equipotential tubes surrounding a line charge; and (b) two-dimensional plot of equipotentials

along the wire. Thus we can draw a two-dimensional plot as shown in Fig. 2.11(b).

If we move a 1 C test charge a small distance in the E field, the small amount of work done is

$$dV = -E\,dr$$

Thus the total amount of work done against the E field in moving the test charge from infinity to a point in the field is

$$\int_0^V dV = -\int_\infty^R \frac{\rho_1}{2\pi\epsilon r}\,dr$$

Unfortunately, solution of this equation results in an infinite potential! (Readers might like to try this for themselves.) The way round this is to set the potential to zero at a distance r, where r tends towards infinity. Thus,

$$\int_0^V dV = -\int_r^R \frac{\rho_1}{2\pi\epsilon r}\,dr \qquad (2.16)$$

as $r \to \infty$. Fortunately, we are more usually concerned with potential differences, and so the dummy variable r, in Equation (2.16), cancels out as shown in the following example.

Example 2.6

A 10 m long wire has a charge of 20 μC along it. Determine the flux density and the electric field strength at a radial distance of 0.5 m from the wire. In addition, find the potential difference between points at 0.5 m and 1.5 m from the wire.

Solution

The line is 10 m long, with a charge of 20 μC. Thus the charge density is

$$\rho_l = \frac{20 \times 10^{-6}}{10}$$

$$= 2 \ \mu C \ m^{-1}$$

Let us now consider a section of the wire with length l metre. The charge on this length is

$$q = 2 \times 10^{-6} \times l$$
$$= 2 \times l \ \mu C$$

We want to find the flux density at a radius of 0.5 m from the wire. The surface area of the Gaussian surface will therefore be

$$\text{area} = 2\pi \times 0.5 \times l \ m^2$$

and so the flux density is

$$D = \frac{2 \times 10^{-6} \times l}{2\pi \times 0.5 \times l} r$$

$$= 6.4 \times 10^{-7} r \quad C \ m^{-2}$$

(It is worth noting that the length of the wire cancels out, and so we could have taken any length we liked.) The electric field strength is given by

$$E = \frac{D}{\epsilon_0}$$

$$= 7.2 \times 10^4 r \quad V \ m^{-1}$$
$$= 72 r \quad kV \ m^{-1}$$

As regards the potential difference, we can use Equation (2.16) to give

$$V_{0.5} = \frac{2 \times 10^{-6}}{2\pi\epsilon} (\ln r - \ln 0.5)$$

and,

$$V_{1.5} = \frac{2 \times 10^{-6}}{2\pi\epsilon} (\ln r - \ln 1.5)$$

with r tending to infinity. Therefore,

$$V_{0.5} - V_{1.5} = \frac{2 \times 10^{-6}}{2\pi\epsilon}\left((\ln r - \ln 0.5) - (\ln r - \ln 1.5)\right)$$

$$= \frac{2 \times 10^{-6}}{2\pi\epsilon}(\ln 1.5 - \ln 0.5)$$

$$= \frac{2 \times 10^{-6}}{2\pi\epsilon}\ln 3$$

$$= 40\,\text{kV}$$

Mathematical approach

In the previous section we used Gauss' law to examine the field around an infinitely long charged line. Here we will split the line into infinitesimally small sections, so that we can use Coulomb's law as applied to point charges.

Let us consider a small section of the line as shown in Fig. 2.12. The charge on this section is $\rho_l \times dz$ and so the flux density at point P is

$$\boldsymbol{D} = \frac{\rho_l \times dz}{4\pi r^2} \tag{2.17}$$

We should note that this is a vector quantity acting at an angle to the axis of the line. By taking the origin as shown in Fig. 2.12, the line stretches from $-\infty$ to $+\infty$. Thus we can see that the line is symmetrical about the origin. Now, $\boldsymbol{D}$ can be split into a radial component, $\boldsymbol{D}_r$, and a vertical component, $\boldsymbol{D}_z$; given by

$$\boldsymbol{D}_r = \frac{\rho_l \times dz}{4\pi r^2} \sin\theta\, \boldsymbol{r} \tag{2.18}$$

and,

$$\boldsymbol{D}_z = \frac{\rho_l \times dz}{4\pi r^2} \cos\theta\, \boldsymbol{z} \tag{2.19}$$

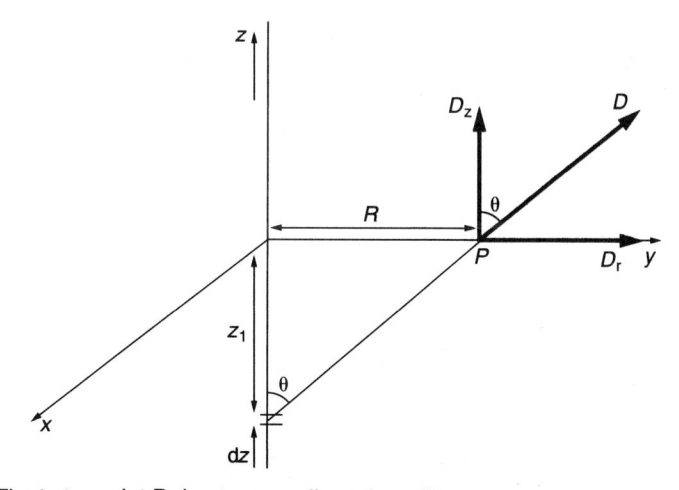

Fig. 2.12 Field at a point P due to a small section of line

28 Electrostatic fields

These flux densities are due to an incremental section of the line. So, to find the total flux density, we can integrate Equations (2.18) and (2.19) with respect to z, between the limits of $-\infty$ and $+\infty$. Unfortunately, as we move up and down the line, r and θ vary as well. If we change variables to integrate with respect to θ, we need to express z and r in terms of θ. So, to return to Equation (2.18) we have,

$$\boldsymbol{D}_r = \frac{\rho_l \times dz}{4\pi r^2} \sin\theta \; \boldsymbol{r}$$

Now,

$$\tan\theta = \frac{R}{z_1}$$

which, after differentiation, becomes

$$\frac{d\theta}{\cos^2\theta} = -\frac{R}{z_1^2} dz$$

and so,

$$dz = -\frac{z_1^2 \, d\theta}{R\cos^2\theta}$$

Thus Equation (2.18) becomes

$$\boldsymbol{D}_r = \frac{\rho_l}{4\pi} \times \frac{\sin\theta}{r^2} \times \frac{-z_1^2}{R\cos^2\theta} d\theta \; \boldsymbol{r}$$

Now, $\cos\theta = z_1/r$ and so we can write

$$\boldsymbol{D}_r = -\frac{\rho_l}{4\pi} \times \sin\theta \times \frac{\cos^2\theta}{R\cos^2\theta} d\theta \; \boldsymbol{r}$$

$$= -\frac{\rho_l}{4\pi R} \sin\theta \, d\theta \; \boldsymbol{r} \qquad (2.20)$$

To find the total radial flux density, we need to integrate this equation with respect to θ. As this is an infinite line, the limits of θ are 0 and π. Thus,

$$\boldsymbol{D}_r = -\frac{\rho_l}{4\pi R} \int_0^\pi \sin\theta \, d\theta \; \boldsymbol{r}$$

$$= -\frac{\rho_l}{4\pi R} \left| -\cos\theta \, d\theta \right|_0^\pi \boldsymbol{r}$$

$$= -\frac{\rho_l}{4\pi R}(-2) \; \boldsymbol{r}$$

$$= -\frac{\rho_l}{2\pi R} \boldsymbol{r} \qquad (2.21)$$

This is exactly the same as the radial field given by Equation (2.14). However, what about the flux density in the z-direction? Well, the line is symmetrical about $z = 0$. Thus there will be an identical component of D_z, acting in the opposite direction, due to an incremental section at $z = -z_1$. Hence we can say that the axial component of D will be zero. (Readers can confirm this for themselves by integrating Equation (2.19) with respect to θ.) As regards the E field, and the potential, we can follow an identical procedure to that used in the previous section.

So, this method has resulted in exactly the same result as that obtained using Gauss' law. Although this derivation has involved us in a considerable amount of work, it has introduced us to the question of symmetry, and the resultant simplifications it can bring.

2.7 SURFACE CHARGES

The last section concentrated on line charges that we find when we have a charged wire. We also come across surface charges, such as those on capacitors and electrostatic precipitators. In such cases we can again use a mathematical approach, or we can apply Gauss' law. In common with the previous section, we will use both approaches.

Gauss' law approach

Let us consider the circular charged plate shown in Fig. 2.13. This plate has a certain charge spread over its surface. To simplify the analysis, let us assume

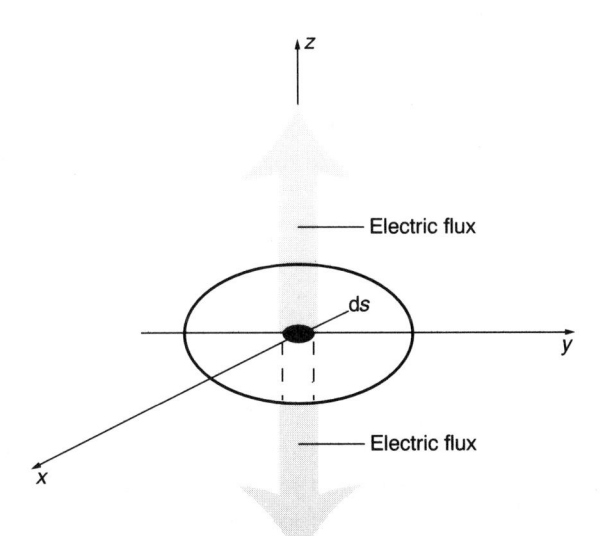

Fig. 2.13 Radiation of electric flux from a charged surface

that the charge distribution is uniform, and that there are no edge effects. Now let us consider a small area of the plate. This area will contain a certain amount of charge, dQ, given by

$$dQ = \rho_s \, ds \tag{2.22}$$

where ρ_s is the surface charge density in $C\,m^{-2}$, and ds is the area of the section. Flux emanating from this area will flow upwards and downwards to occupy a cylinder. (There will only be a vertical component of flux because any horizontal flux will cancel out due to symmetry. This is a similar situation to that which we met when we examined line charges.) By applying Gauss' law, we see that the total flux out of the cylinder, in both directions, must equal the enclosed charge. So, half the flux flows upwards, and half flows downwards. Thus the flux density at any height above the disc is

$$\boldsymbol{D}_z = \frac{dQ}{2\,ds}\,\boldsymbol{z}$$

$$= \frac{\rho_s \, ds}{2\,ds}\,\boldsymbol{z}$$

$$= \frac{\rho_s}{2}\,\boldsymbol{z} \tag{2.23}$$

and the electric field strength is

$$\boldsymbol{E}_z = \frac{\rho_s}{2\epsilon_0\epsilon_r}\,\boldsymbol{z} \tag{2.24}$$

The important thing to note here is that the $\boldsymbol{E}$ field is independent of the distance from the disc. This is a consequence of having the flux flow in a cylindrical tube.

As regards the potential, we have equipotential surfaces parallel to the disc. These surfaces will have the same area as the disc. If we take zero potential at infinity, the absolute potential at a distance z from the disc will be

$$\int_0^V dV = \int_\infty^z \frac{\rho_s}{2\epsilon_0\epsilon_r}\,dz$$

Therefore,

$$V = \frac{\rho_s}{2\epsilon_0\epsilon_r}\,z \tag{2.25}$$

This is an example of a linear field because the $\boldsymbol{E}$ field is constant regardless of the distance from the disc. We should remember, however, that this result has only appeared because we have a uniform charge density.

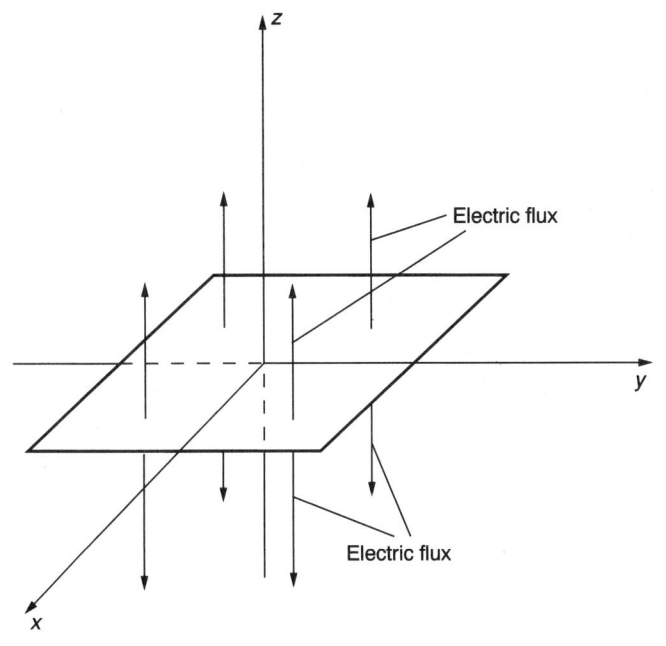

Fig. 2.14 Radiation of electric flux from a charged square

Example 2.7

Determine the flux density, and hence the electric field strength, produced at a distance of 1 metre from the centre of a 1 m² square of insulating material that has a total charge of 10 pC evenly distributed over it. Assume that the square is in air.

Solution

Let us place the square at the centre of a set of Cartesian axes as shown in Fig. 2.14. The square lies in the xy-plane, and so the electric flux will act equally along the positive and negative z-axis. As we have just seen, the electric flux density is independent of the exact shape of the material, and is given by Equation (2.23) as

$$D_z = \frac{\rho_s}{2} z$$

Thus,

$$D_z = \frac{10 \times 10^{-12}}{2} z$$

$$= 5z \ \text{pC m}^{-2}$$

As regards the E field, we can use $D = \epsilon E$ to give

$$E_z = \frac{5 \times 10^{-12}}{8.854 \times 10^{-12}} z$$

$$= 0.56 \, z \text{ V m}^{-1}$$

Mathematical approach

As we have previously seen, we can only apply Coulomb's law, and hence use our usual expressions for D and E, when considering point charges. However, we have a surface charge, and so how can we analyse this situation? The solution is to consider a small section of the disc, calculate the flux density due to the charge on this small section, and integrate the result over the area of the disc. (Problem 1.2 gives more information about the integration method used in the following derivation.)

As Fig. 2.15 shows, let us consider a small section of a disc. The area of this small section is

$$ds = r \, d\phi \, dr \qquad (2.26)$$

if $d\phi$ is expressed in radians. (This is a direct consequence of expressing $d\phi$ in radians. The circumference of the disc is $2\pi r$, and this encloses an angle of 360°, or 2π radians. So, the length of an arc that subtends an angle of 180°, or π radians, is πr. Thus, the length of an arc is equal to the product of the angle (in radians) and the radius of the arc.)

If we have a charge spread over the surface of this disc, the charge on this small section is

$$dQ = \rho_s \, ds \qquad (2.27)$$

which we can take to be a point charge if ds is very small. This charge will produce a small component of the flux density, acting in the direction shown in Fig. 2.15(a). So,

$$dD = \frac{\rho_s \, ds}{4\pi l^2} \qquad (2.28)$$

We can resolve this flux density into horizontal and vertical components, and then integrate with respect to ϕ between the limits 0 and 2π. This integration will describe a ring of thickness dr, and radius r. It is then a matter of integrating with respect to r to map out the whole of the disc. However, when we integrate with respect to ϕ, we find that the **horizontal** component of D will be zero. This is a consequence of the symmetry of the disc, similar to the symmetry we met in the previous section. (Readers can check this by performing the integration for themselves.)

So, because of symmetry, we only need to consider the vertical component of the flux density. Thus,

2.7 Surface charges

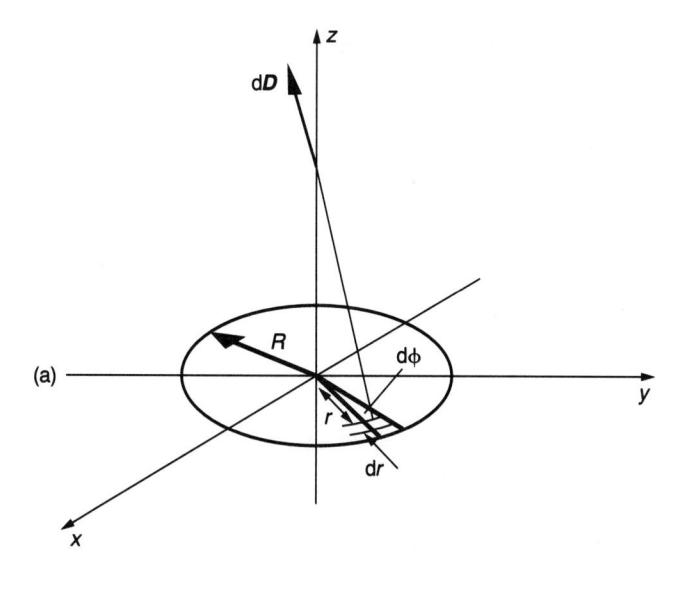

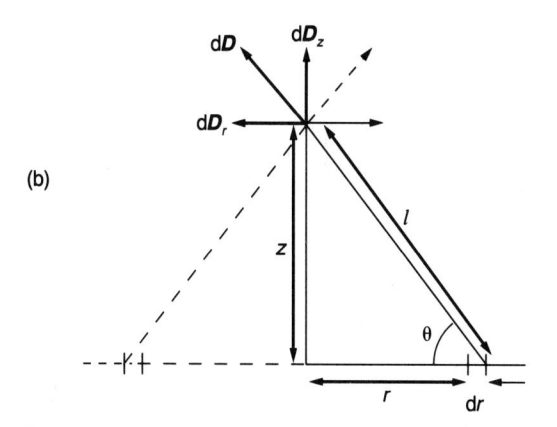

Fig. 2.15 (a) Field at point P due to a small section of a charged disc; and (b) side view

$$dD_z = \frac{\rho_s \, ds}{4\pi l^2} \sin\theta \, \mathbf{z}$$

$$= \frac{\rho_s \, ds}{4\pi l^2} \frac{z}{l} \mathbf{z}$$

Now, from Equation (2.26), we can write

$$dD_z = \frac{\rho_s r \, d\phi \, dr}{4\pi l^2} \frac{z}{l} \mathbf{z} \tag{2.29}$$

The total flux density is obtained by integrating this equation with respect to ϕ and r. We can perform integration with respect to ϕ very easily because l does not vary as we move in a circular direction. (The integration of ds with respect

34 Electrostatic fields

to ϕ describes a ring of radius r.) So, the flux density due to a ring of thickness dr and radius r is

$$d\boldsymbol{D}_z = \frac{\rho_s r\, dr\, z}{4\pi l^3} \int_0^{2\pi} d\phi\, \boldsymbol{z}$$

$$= \frac{\rho_s r\, dr\, z}{4\pi l^3} 2\pi\, \boldsymbol{z}$$

$$= \frac{\rho_s r\, dr\, z}{2 l^3}\, \boldsymbol{z} \qquad (2.30)$$

We now need to integrate Equation (2.30) with respect to radius to find the total flux density at the point P. Unfortunately, as we perform the integration, the length l varies as the radius goes from 0 to R. Thus, we need to express l in terms of r prior to integrating with respect to r. So,

$$\boldsymbol{D}_z = \frac{\rho_s z}{2} \int_0^R \frac{r\, dr}{l^3}\, \boldsymbol{z}$$

$$= \frac{\rho_s z}{2} \int_0^R \frac{r\, dr}{(r^2 + z^2)^{3/2}}\, \boldsymbol{z}$$

$$= \frac{\rho_s z}{2} \left| \frac{-1}{(r^2 + z^2)^{1/2}} \right|_0^R \boldsymbol{z}$$

i.e.

$$\boldsymbol{D}_z = \frac{\rho_s z}{2} \left(\frac{1}{z} - \frac{1}{(R^2 + z^2)^{1/2}} \right) \boldsymbol{z} \qquad (2.31)$$

We can perform a very simple check on this equation by letting $z \to \infty$ (i.e. find the flux density at infinity). Under these circumstances, we would expect the disc to approximate to a point charge, and so the flux density should approximate to Equation (2.4). So, by using the binomial expansion of the term in brackets, we get

$$\boldsymbol{D}_z \approx \frac{\rho_s z}{2} \left(\frac{1}{z} - \frac{1}{z} + \frac{1}{2z} \frac{R^2}{z^2} \right) \boldsymbol{z}$$

$$= \frac{\rho_s z}{2} \frac{1}{2z} \frac{R^2}{z^2}\, \boldsymbol{z}$$

$$= \rho_s \frac{R^2}{4z^2}\, \boldsymbol{z}$$

$$= \frac{Q}{\pi R^2} \frac{R^2}{4z^2} \mathbf{z}$$

$$= \frac{Q}{4\pi z^2} \mathbf{z}$$

which is the equation for the flux density resulting from a point charge.

We can perform another simple check, by letting $R \to \infty$. If we do this, we find, from Equation (2.31), that

$$\mathbf{D}_z \to \frac{\rho_s}{2} \mathbf{z}$$

which is the same as that found using Gauss' law, Equation (2.23).

So, we have successfully shown that the flux density from a surface charge distribution reduces to that of a point charge, if the distance from the surface is large enough. We have also shown that the mathematical approach gives the same result as that obtained using Gauss' law.

2.8 VOLUME CHARGES

The previous two sections have introduced us to line and surface charge densities. However, we live in a three-dimensional world (four if you count time) and so we will often meet charged volumes. When considering a volume charge density, we can make good use of Gauss' law to replace the volume charge by a point charge at the centre of the volume.

As an example, let us consider a sphere with a charge evenly distributed throughout its volume, as shown in Fig. 2.16(a). To analyse this situation, we

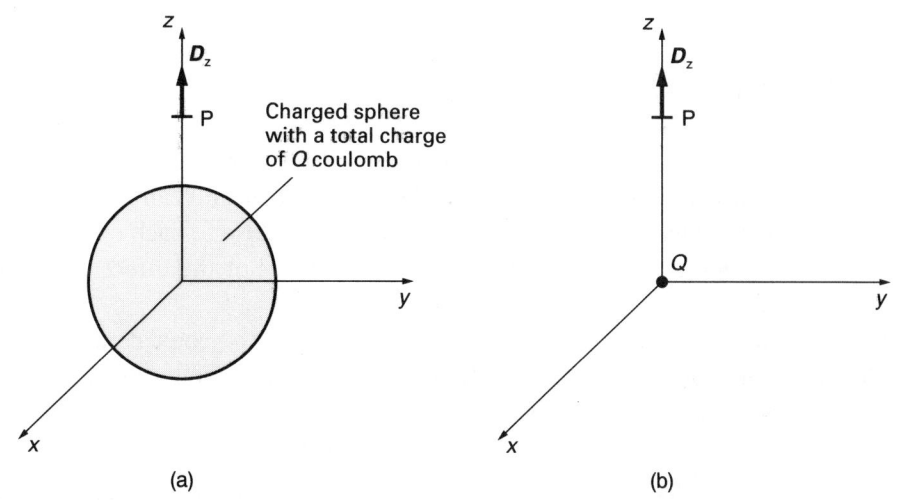

Fig. 2.16 (a) Field at point P due to a charged sphere; and (b) simplification due to application of Gauss' law

could consider a small section of the sphere, and perform an integration to map out the whole of the volume. However, this will involve us in a considerable amount of work. An alternative is to apply Gauss' law, and replace the volume charge density by a point charge at the centre of the volume.

So, if the sphere has a total charge of Q coulomb distributed throughout the volume, we can replace the sphere by a point charge, placed at the centre of the sphere, of magnitude Q. This is shown in Fig. 2.16(b). It is then a simple matter to find the flux density, etc., at any distance from the surface of the sphere.

Example 2.8

A solid sphere, of radius 0.25 m, has a charge of 10 pC evenly distributed throughout its volume. By using Gauss' law, determine the flux density at a point 1 m from the centre of the sphere.

Solution

We have to apply Gauss' law to the sphere. Now, the total charge contained throughout the sphere is 10 pC. So, we can replace the sphere by a point charge of 10 pC placed at the centre of the sphere. We now need to determine the flux density at a distance of 1 m from this charge. Thus,

$$D_r = \frac{10 \times 10^{-12}}{4\pi 1^2} r$$

$$= 0.8 \, r \quad \text{pC m}^{-2}$$

Although we could have considered the flux due to a small incremental volume of the sphere, and then integrated throughout the volume of the sphere, the mathematics would be very complicated indeed. (Interested readers can show this for themselves, but it is not recommended!)

2.9 CAPACITORS

Most of us are familiar with capacitors as circuit elements and, as such, we seldom need to examine their structure. What is not often realized is that a capacitor is formed whenever we have two conductors close to each other. Such a situation frequently occurs in electrical engineering, but the effect does not make itself felt until we reach high frequencies.

In this section we will consider parallel plate capacitors, coaxial cable, twin feeder, and microstrip line. All of these capacitors are commonly found in electrical engineering. We begin our study by examining the parallel plate capacitor.

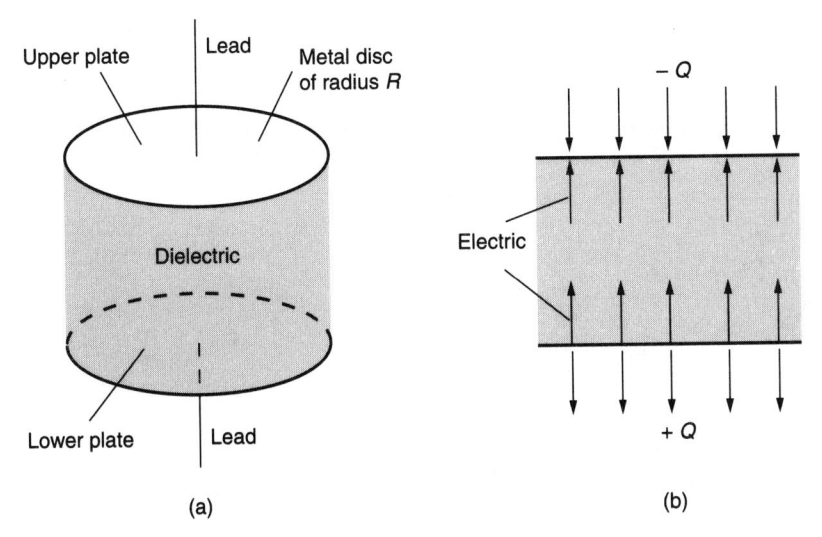

Fig. 2.17 (a) Basic structure of a parallel plate capacitor; and (b) flux distribution in a parallel plate capacitor

Parallel plate capacitors

Capacitors come in a variety of shapes and sizes: from big electrolytic capacitors for smoothing the output of power supplies, to small-value disc ceramic capacitors for use in high-frequency circuits. All types of capacitor are based on the simple structure shown in Fig. 2.17.

For convenience, we will consider a circular plate capacitor with R being the radius of each plate. When the bottom plate of the capacitor has a charge on it, this charge induces an equal and opposite charge in the top plate. Thus, if the bottom electrode has a charge of $+Q$ on it, the charge on the upper plate is $-Q$.

Now, as we saw in Section 2.7, if the charge is evenly distributed over the plate, the flux is equally divided into upward and downward flux. So, flux flows upwards through the dielectric from the positively charged lower plate. As we have an upper plate with a charge of $-Q$ on it, flux in the upper half of the capacitor will flow upwards. Thus the total flux in the capacitor flows in the positive z-direction as shown in Fig. 2.17(b).

As we have two equal sources of flux acting in the capacitor, the total flux density in the capacitor is

$$\boldsymbol{D}_z = \left(\frac{\rho_s}{2} + \frac{\rho_s}{2}\right)\boldsymbol{z}$$

$$= \frac{Q}{\pi R^2}\boldsymbol{z} \qquad (2.32)$$

and so the electric field strength is

$$E_z = \frac{Q}{\epsilon_0 \epsilon_r \pi R^2} z \qquad (2.33)$$

Thus the potential difference between the top and bottom plates is

$$\int_{V_1}^{V_2} dV = \frac{-Q}{\epsilon_0 \epsilon_r \pi R^2} \int_{-d/2}^{d/2} -dz$$

and so,

$$V_2 - V_1 = \frac{Q}{\epsilon_0 \epsilon_r \pi R^2} \left(\frac{d}{2} + \frac{d}{2} \right)$$

$$= \frac{Q}{\epsilon_0 \epsilon_r \pi} \frac{d}{R^2}$$

or, more generally,

$$V = \frac{Q}{\epsilon_0 \epsilon_r \pi R^2} d$$

We can rearrange this equation to give

$$Q = \frac{\epsilon_0 \epsilon_r \pi R^2}{d} V$$

which shows that the stored charge is directly proportional to the voltage across the capacitor plates. The constant of proportionality is the capacitance given by

$$C = \frac{\epsilon_0 \epsilon_r \times \text{area}}{d} \qquad (2.34)$$

and so we can also write

$$Q = CV \qquad (2.35)$$

Both equations should be familiar to most readers.

Coaxial cable

Coaxial cable is very widely used in everyday life: a familiar example is the lead that connects the aerial to the television set, or the patch lead between a video player and the television. Figure 2.18 shows the basic structure of this type of cable.

Under normal circumstances, the inner conductor of the cable carries the signal, or voltage, whereas the outer conductor is usually earthed. The advantage of this structure is that any external interference has to pass through an earthed conductor before it reaches the signal. In effect, the outer conductor shields the signal from any external interference.

2.9 Capacitors

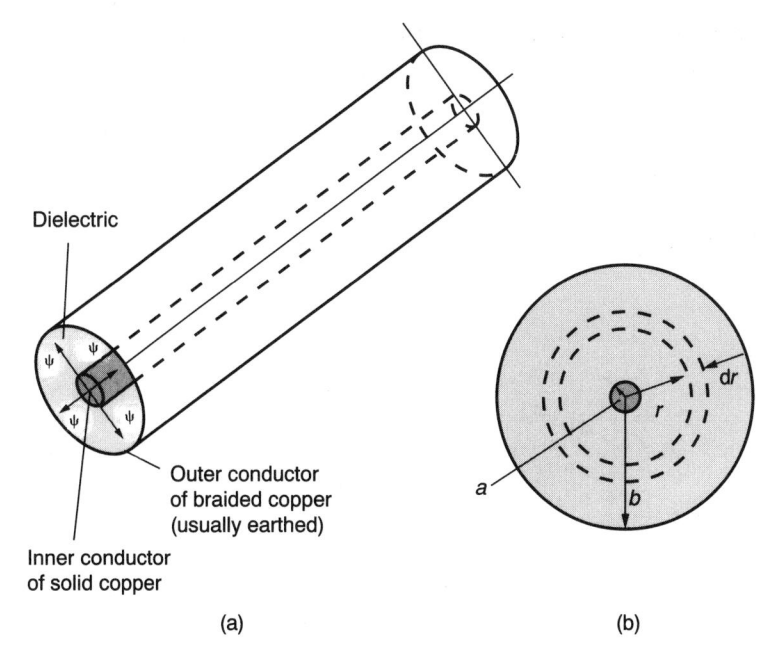

Fig. 2.18 (a) Basic structure of coaxial cable; and (b) end view of coaxial cable

Now, if the inner conductor is at a certain potential above the earthed shield, we will have a capacitor. To find the capacitance, we need an equation that links the potential difference between inner and outer conductors to the charge on the inner conductor.

As we are dealing with a length of cable, let us assume that the inner conductor has a charge of ρ_l coulomb per unit length. This charge will produce flux in a radial direction similar to that which we met in Section 2.6. The Gaussian surface in this instance is a tube of radius r, thickness dr, and length l. So, the flux flowing through this surface is

$$\psi = \rho_l \times l$$

As this flux flows in a radial direction, the flux density through the Gaussian surface is

$$\boldsymbol{D} = \frac{\rho_l l}{2\pi r l} \boldsymbol{r}$$

or,

$$\boldsymbol{D}_r = \frac{\rho_l}{2\pi r} \boldsymbol{r}$$

and so the electric field strength at this radius is

$$\boldsymbol{E}_r = \frac{\rho_l}{2\pi \epsilon_0 \epsilon_r r} \boldsymbol{r}$$

Now, the thickness of the Gaussian surface is dr, and so the potential difference across the surface is

$$dV = -E_r \, dr$$

$$= -\frac{\rho_1}{2\pi\epsilon_0\epsilon_r} \frac{1}{r} dr$$

The potential difference between the inner and outer conductors is therefore given by

$$\int_0^V dV = -\frac{\rho_1}{2\pi\epsilon_0\epsilon_r} \int_b^a \frac{1}{r} dr$$

and so

$$V = -\frac{\rho_1}{2\pi\epsilon_0\epsilon_r} \Big|\ln r\Big|_b^a$$

$$= \frac{\rho_1}{2\pi\epsilon_0\epsilon_r} \ln\left(\frac{b}{a}\right)$$

$$= \frac{Q}{2\pi\epsilon_0\epsilon_r l} \ln\left(\frac{b}{a}\right)$$

Thus the capacitance is

$$C = \frac{Q}{V}$$

$$= \frac{2\pi\epsilon_0\epsilon_r}{\ln(b/a)} \times \text{length} \qquad (2.36)$$

and so the capacitance per unit length is

$$C' = \frac{2\pi\epsilon_0\epsilon_r}{\ln(b/a)} \; \text{F m}^{-1} \qquad (2.37)$$

We should note that the capacitance per unit length is directly dependent on the length of the cable. So, if we double the length of the cable, the capacitance also doubles.

Example 2.9

A 500 m length of coaxial cable has an inner conductor of radius 2 mm, and an outer conductor of radius 1 cm. The relative permittivity of the dielectric separating the inner and outer conductors is 5. Determine the capacitance of the cable. If the inner conductor is at a potential of 1 kV above that of the outer conductor, determine the maximum value of the E field in the dielectric.

Solution

We want to find the capacitance of the cable. So, we can use Equation (2.36) to give

$$C = \frac{2\pi\epsilon_0\epsilon_r}{\ln(b/a)} \times \text{length}$$

$$= \frac{2\pi \times 8.854 \times 10^{-12} \times 5}{\ln(10^{-2}/2 \times 10^{-3})} \times 500$$

and so,

$C = 86.4 \text{ nF}$

or,

$C' = 173 \text{ pF m}^{-1}$

As regards the **E** field in the dielectric, we have seen that

$$\boldsymbol{E}_r = \frac{\rho_l}{2\pi\epsilon_0\epsilon_r r}\boldsymbol{r}$$

In order to find $\boldsymbol{E}_r$, we need to know the charge per unit length, ρ_l. As $Q = CV$, Equation (2.35), we can write

$\rho_l = C'V$
$= 173 \times 10^{-12} \times 1 \times 10^3$
$= 173 \text{ nC m}^{-1}$

As the **E** field is inversely proportional to radius, the maximum field occurs at the surface of the inner conductor. Thus,

$$E_r|_{max} = \frac{173 \times 10^{-9}}{2\pi \times 8.854 \times 10^{-12} \times 5 \times 2 \times 10^{-3}}$$

$= 311 \text{ kV m}^{-1}$

This is a quite considerable field strength, and one that may cause the dielectric to break down. Chapter 6 deals with this in greater detail.

Twin feeder

In communications, twin feeder is often used as connecting wire between short-wave transmitters and their aerials. We also find twin feeder in power transmission systems and telephone lines. As Fig. 2.19 shows, twin feeder generally consists of two parallel wires held apart by some means.

Let us assume that the left-hand conductor has a charge per unit length of ρ_l C m^{-1}. This charge will induce an equal and opposite charge on the right-hand conductor. The flux emanating from the left-hand conductor does so in a

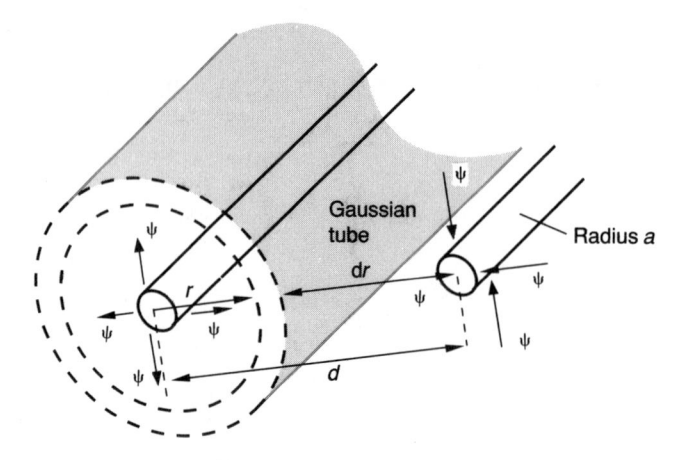

Fig. 2.19 Basic structure of twin feeder

radial fashion, and so we can take a Gaussian tube of length l, similar to that which we used in Section 2.6.

The charge on a length of the left-hand conductor is $\rho_l \times l$ coulomb. Thus the total flux through the Gaussian surface is

$$\psi = \rho_l \times l \tag{2.38}$$

The area of the Gaussian surface is $2\pi r \times l$ and so the flux density at a radius r is

$$\boldsymbol{D}_r = \frac{\rho_l \times l}{2\pi r \times l} \boldsymbol{r}$$

$$= \frac{\rho_l}{2\pi r} \boldsymbol{r} \tag{2.39}$$

Thus the electric field strength at this radius is

$$\boldsymbol{E}_r = \frac{\rho_l}{2\pi \epsilon_0 r} \boldsymbol{r} \tag{2.40}$$

We also have a right-hand conductor with an equal and opposite charge. The radial electric field from this conductor will have the same direction as the field due to the left-hand conductor. The electric field strength due to the right-hand conductor is

$$\boldsymbol{E}_r = \frac{\rho_l}{2\pi \epsilon_0 (d-r)} \boldsymbol{r} \tag{2.41}$$

and so the total electric field strength is

$$\boldsymbol{E}_r = \frac{\rho_l}{2\pi \epsilon_0 r} + \frac{\rho_l}{2\pi \epsilon_0 (d-r)} \boldsymbol{r} \tag{2.42}$$

2.9 Capacitors

To find the potential between the two lines, we must integrate this equation with respect to radius. So,

$$\int_{V_b}^{V_a} dV = \frac{-\rho_1}{2\pi\epsilon_0} \int_{d-a}^{a} \left(\frac{1}{r} + \frac{1}{(d-r)}\right) dr$$

Therefore,

$$V_a - V_b = \frac{-\rho_1}{2\pi\epsilon_0} \left|\ln r - \ln(d-r)\right|_{d-a}^{a}$$

$$= \frac{-\rho_1}{2\pi\epsilon_0}(\ln a - \ln(d-a) + \ln a - \ln(d-a))$$

$$= \frac{-\rho_1}{\pi\epsilon_0}(\ln a - \ln(d-a))$$

$$= \frac{\rho_1}{2\pi\epsilon_0}(\ln(d-a) + \ln a)$$

$$= \frac{\rho_1}{\pi\epsilon_0} \ln\left(\frac{d-a}{a}\right)$$

$$= \frac{Q}{\pi\epsilon_0} \ln\left(\frac{d-a}{a}\right) \times \text{length} \quad (2.43)$$

So, the capacitance of the arrangement is

$$C = \frac{\pi\epsilon_0}{\ln|(d-a)/a|} \times \text{length F}$$

or,

$$C' = \frac{\pi\epsilon_0}{\ln|(d-a)/a|} \text{ F m}^{-1} \quad (2.44)$$

If the separation of the conductors is significantly greater than the diameter of the conductors, Equation (2.44) reduces to

$$C' = \frac{\pi\epsilon_0}{\ln(d/a)} \text{ F m}^{-1} \quad (2.45)$$

Example 2.10

A 200 m length of feeder consists of two 2 mm radius conductors separated by a distance of 20 cm. Determine the capacitance of the arrangement.

Solution

As the distance between the conductors is very much greater than the radius of the conductors, we can use Equation (2.45) to give

$$C' = \frac{\pi \epsilon_0}{\ln(d/a)}$$

$$= \frac{\pi \times 8.854 \times 10^{-12}}{\ln(20 \times 10^{-2}/2 \times 10^{-3})}$$

$$= 6 \text{ pF m}^{-1}$$
$$= 1.2 \text{ nF} \quad \text{for the 200 m length}$$

Wire over ground – the method of images

In the last section, we studied twin feeder. However, we often come across conductors placed over a ground-plane. Under these circumstances, we can make use of the method of images to find the capacitance of the arrangement.

Figure 2.20(a) shows the situation we are considering. As we can see, we have a line above an infinite ground-plane. (A ground-plane is simply a large conducting area that is earthed. The Earth itself is one example. Another example is the use of double-sided printed circuit boards.) To analyse this situation, we can introduce an imaginary conductor on the other side of the ground-plane, as shown in Fig. 2.20(b).

As the ground-plane is exactly half-way between the two conductors, it is effectively lying along an equipotential line. This means that we can remove the ground-plane, so leaving us with the twin feeder we have just considered. As we can see from Fig. 2.20(c), we can consider the two-wire situation as being made up of two single-wire/ground-plane arrangements in series. Thus the capacitance of the single conductor over a ground-plane is simply twice that of the twin feeder. We can now write

$$C' = \frac{2\pi\epsilon_0}{\ln((d/2 - a)/a)} \text{ F m}^{-1} \tag{2.46}$$

or,

$$C' = \frac{2\pi\epsilon_0}{\ln(d/2a)} \text{ F m}^{-1} \tag{2.47}$$

where $d/2$ is the height of the conductor above the ground-plane.

Example 2.11

A high-voltage power line consists of 1 cm radius copper wire placed 25 m above the ground. Determine the capacitance that the line has to the ground.

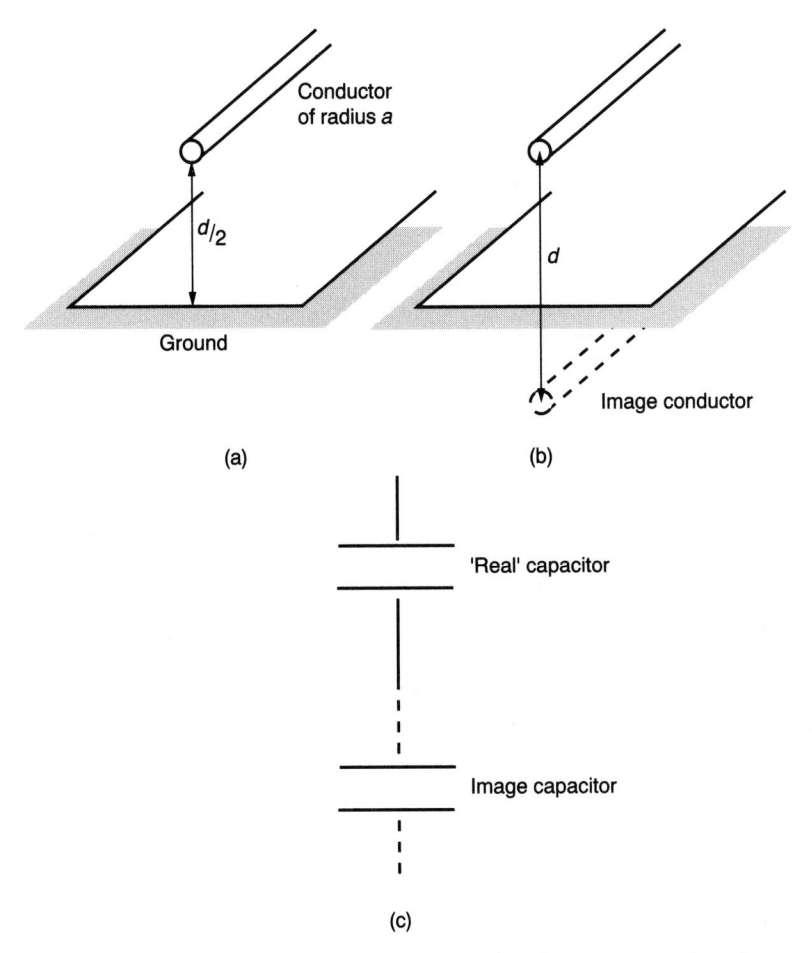

Fig. 2.20 (a) Wire over ground; (b) image conductor for wire over ground; and (c) equivalent circuit of wire and image

Solution

The height of the wire above the ground is significantly greater than the radius of the wire, and so we can use Equation (2.47) to give

$$C' = \frac{2\pi\epsilon_0}{\ln(d/2a)} \text{ F m}^{-1}$$

$$= \frac{2\pi \times 8.854 \times 10^{-12}}{\ln(25/2 \times 10^{-12})}$$

$$= 7.8 \text{ pF m}^{-1}$$

Microstrip lines

When constructing electronic circuits, we often use double-sided printed circuit board. When using this type of board, the upper layer generally carries the

46 Electrostatic fields

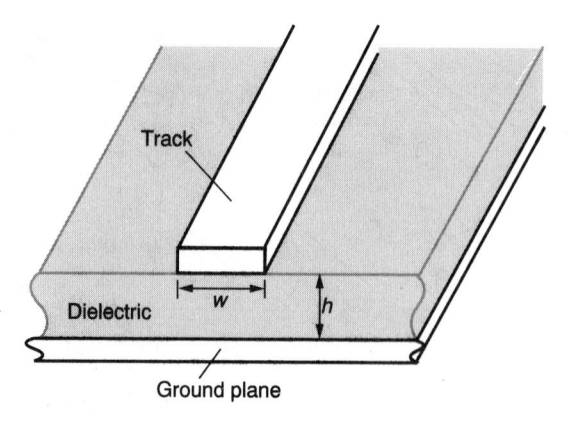

Fig. 2.21 Cross-section through a double-sided printed circuit board

signal, while the bottom layer is usually earthed. Under these circumstances, the signal line and earth form a capacitor. When operating at low frequencies, the effect is not very pronounced. However, at high frequencies, the capacitance has a greater effect.

Figure 2.21 shows the cross-section through a double-sided circuit board. In common with the previous example, we can make good use of the method of images to produce a parallel plate capacitor. Thus the capacitance of the microstrip will be

$$C = \frac{2\epsilon_0\epsilon_r \times \text{area}}{2h}$$

or,

$$C' = \frac{\epsilon_0\epsilon_r w}{h} \text{ F m}^{-1} \tag{2.48}$$

If the width of the track is much smaller than the thickness of the board, we can approximate the distribution to that of a cylindrical wire over a ground-plane. This is identical to the situation we met in the last section. Thus, the capacitance per unit length is

$$C' = \frac{2\pi\epsilon_0\epsilon_r}{\ln(h/w)} \text{ F m}^{-1} \tag{2.49}$$

Example 2.12

A 3 mm wide track is etched on one side of some double-sided printed circuit board. The thickness of the board is 2 mm, and the dielectric has a relative permittivity of 5. Determine the capacitance per cm.

Solution

As the track width is of the same order of magnitude as the board thickness, we must use Equation (2.48) to give

$$C' = \frac{\epsilon_0 \epsilon_r w}{h} \text{ F m}^{-1}$$

$$= \frac{8.854 \times 10^{-12} \times 5 \times 3 \times 10^{-3}}{2 \times 10^{-3}} \text{ F m}^{-1}$$

$$= 66 \text{ pF m}^{-1}$$
$$= 0.66 \text{ pF cm}^{-1}$$

Energy storage

We can use capacitors as energy storage devices – on computer memory boards, charged capacitors can supply power to the memory chips if the main supply fails. In addition, anyone who has worked on the high-voltage part of a television will know that a capacitor stores energy that must be discharged before working on the chassis. So, we can use a capacitor to store energy, but how and where does a capacitor store the energy?

As we have already seen, if the capacitor is holding a charge an electric field exists in the dielectric. When the capacitor discharges through an external circuit, charges appear to move across the dielectric, against the E field. As they move against the field, work is done, and energy is lost. When a discharged capacitor is connected to a voltage source, the reverse takes place.

To find the stored energy, let us take a capacitor connected to a source of V volts. If we increase the voltage by dV, the stored charge will increase by dQ. If these changes occur in a time dt, the instantaneous current will be

$$i = \frac{dQ}{dt}$$

$$= C \frac{dV}{dt} \tag{2.50}$$

The instantaneous power is given by

$$iV = CV \frac{dV}{dt}$$

and so the energy supplied in raising the voltage from V to $V + dV$ in time dt is

$$iV \, dt = CV \frac{dV}{dt} \times dt$$

or,

energy = $CV\,dV$

Thus the total energy supplied in raising the capacitor voltage from zero to V is

$$\text{energy} = \int_0^V CV\,dV$$

$$= C\left|\frac{V^2}{2}\right|_0^V$$

$$= \tfrac{1}{2}CV^2 \text{ joule} \qquad (2.51)$$

Let us now find where the energy is stored. If we substitute for the capacitance in Equation (2.51) we get

$$\text{energy} = \frac{1}{2}\frac{\epsilon_0 \epsilon_r \times \text{area}}{d}V^2$$

$$= \tfrac{1}{2}\epsilon_0\epsilon_r \frac{V^2}{d} \times \text{area}$$

$$= \tfrac{1}{2}\epsilon_0\epsilon_r\, E\, V \times \text{area} \qquad (2.52)$$

We can find the energy per unit volume by dividing this equation by the volume of the capacitor. So,

$$\text{energy} = \tfrac{1}{2}\epsilon_0\epsilon_r\, E\, \frac{V}{d} \times \frac{\text{area}}{\text{area}}$$

$$= \tfrac{1}{2}DE\ \text{J m}^{-3} \qquad (2.53)$$

So, from Equation (2.53) it would appear that the electrostatic field stores the energy and not the capacitor plates. Such a point of view is quite useful if we consider fields in free space.

Example 2.13

A 10 µF capacitor has a potential difference between the plates of 50 V. Determine the energy stored in the electrostatic field.

Solution

As the capacitance is quoted, we can use Equation (2.51) to give

energy = $\tfrac{1}{2}CV^2$
 = $\tfrac{1}{2} \times 10 \times 10^{-6} \times 50^2$
 = 12.5 mJ

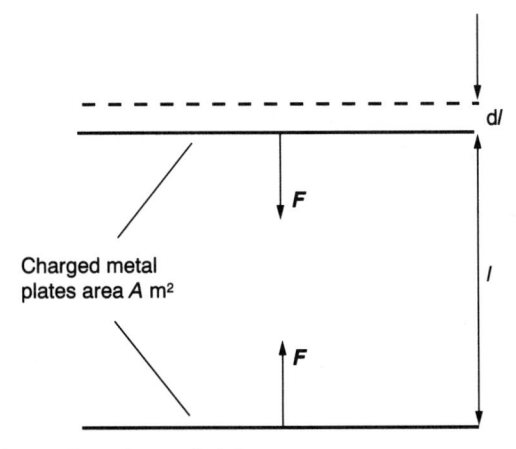

Fig. 2.22 Force between two charged plates

Force between charged plates

We have already seen that charges exert a force on each other. We have also seen that charged circular plates radiate electric flux in a cylinder. We should therefore expect that two charged plates will exert a force on each other. To find this force, we could adapt our model of flux distribution in a parallel plate capacitor, and then calculate the force between the plates. However, there is a simpler method.

Let us consider the parallel plate capacitor shown in Fig. 2.22. This capacitor stores a certain amount of energy given by

energy $= \frac{1}{2} DE \times$ area $\times l$

Now, there will be a force of attraction between the two plates. If we move the top plate by a small amount dl, we do work against the attractive force. As we are moving the top plate a very small amount, the E field will hardly vary. As the work done must equal the change in stored energy, we can write,

$$F\, \mathrm{d}l = \tfrac{1}{2} DE \times \text{area} \times (l + \mathrm{d}l) - \tfrac{1}{2} DE \times \text{area} \times l$$
$$= \tfrac{1}{2} DE \times \text{area} \times \mathrm{d}l \qquad (2.54)$$

As D is the flux density, Equation (2.54) becomes,

$F = \tfrac{1}{2} \psi E$

or,

$F = \tfrac{1}{2} QE$ newton (2.55)

where we have made use of Gauss' law. Depending on the particular application, the force between the two plates can be quite high.

Example 2.14

An electrostatic voltmeter consists of two square plates, of area 25 cm², separated by a distance of 2 cm in air. One of the plates is fixed, while the other is attached to a spring mechanism that deflects a needle in front of a calibrated scale. The constant of proportionality for the meter is 10° per 1 µN of force. Determine the angular displacement of the needle if a potential of 500 V is maintained across the plates.

Solution

The force between the two plates is given by Equation (2.55) as

$F = \frac{1}{2} QE$ newton

Thus,

$F = \frac{1}{2} \times \text{capacitance} \times 500 \times \dfrac{500}{2 \times 10^{-2}}$

$= 7 \times 10^{-6}$ N

This corresponds to an angular displacement of 70°.

Low-frequency effects and displacement current

So far we have only considered the effects of direct current (d.c.) on capacitors. However, in electrical engineering we usually find capacitors in alternating current (a.c.) circuits. So, what effect does a capacitor have on a.c. signals?

Figure 2.23 shows a capacitor connected to an a.c. source. The voltage across the capacitor varies with time and so, if we assume the source to be sinusoidal, we can write

$$v_s(t) = V_{pk} \sin \omega t \tag{2.56}$$

where V_{pk} is the peak source voltage, and ω is the angular frequency of the source.

Now, the capacitance is defined as the ratio of charge to potential difference between the capacitor plates, i.e.

$C = \dfrac{Q}{V}$

or,

$$Q = CV \tag{2.57}$$

As the source is varying with time, we can write

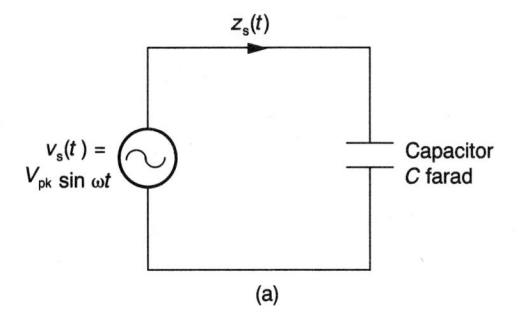

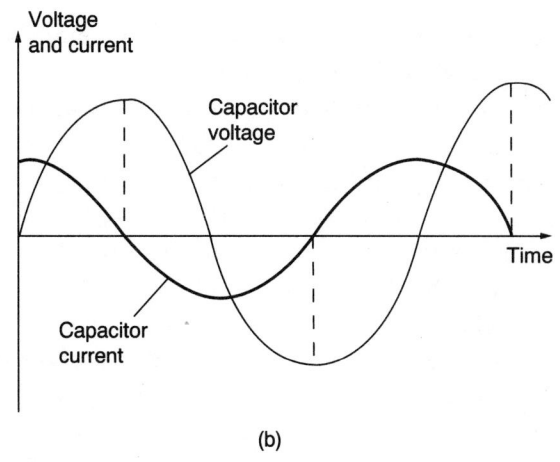

Fig. 2.23 (a) Capacitor connected to an a.c. source; and (b) relationship between capacitor voltage and current

$$q(t) = Cv_s(t)$$
$$= CV_{pk} \sin \omega t$$

If we differentiate this equation with respect to time, we get

$$\frac{d}{dt}q(t) = CV_{pk} \omega \cos \omega t$$

As current is the rate of change of charge with respect to time, we get

$$i_s(t) = \frac{d}{dt}q(t)$$
$$= CV_{pk} \omega \cos \omega t$$
$$= CV_{pk} \omega \sin(\omega t + 90°) \qquad (2.58)$$

So, when connected to an alternating source, the capacitor allows a current to flow, with the current leading the voltage by 90°. (Figure 2.23(b) shows the relationship between capacitor voltage and current.) We should note that the current flow is directly proportional to the angular frequency of the source, i.e. the higher the frequency, the larger the current flow. We can formalize this observation by defining the reactance of the capacitance, X_C, as

$$X_C = \frac{1}{\omega C}$$

$$= \frac{1}{2\pi f C} \tag{2.59}$$

By combining this result with Equation (2.58) we get, after some rearranging,

$$i_s(t) = \frac{v_s(t) \,\underline{/90°}}{X_C}$$

This is remarkably similar to Ohm's law, except that we are dealing with a.c. quantities, and there is a 90° phase shift involved.

So, when a capacitor is connected to an a.c. source, it provides a low resistance path for a.c. signals. Of course, if there is a d.c. voltage across the capacitor, no current will flow (if the capacitor is ideal). This makes capacitors very useful in smoothing power rails (where there might be some variation in supply voltage) and in connecting a.c. amplifier stages together – capacitors will let the a.c. signal through, but block any d.c. levels.

Although this 'circuits' model indicates that a current will flow through a capacitor connected to an a.c. supply, it does not explain how the current gets from one plate to the other. Indeed, if the capacitor dielectric is ideal, there can be no flow of electrons from one plate to the other. So, how does the current magically cross the dielectric? To explain this we must use a field theory approach.

Figure 2.24(a) shows the state of the capacitor on positive half-cycles of the supply voltage. As can be seen, positive charge has built up on the lower plate. As we saw earlier in this section, these charges will radiate flux in upward and downward directions. The flux that radiates upwards will tend to attract negative charges to the top plate, and repel any positive charges. Thus, in the positive half-cycle of the supply voltage, positive charges on the lower plate induce negative charges on the upper plate.

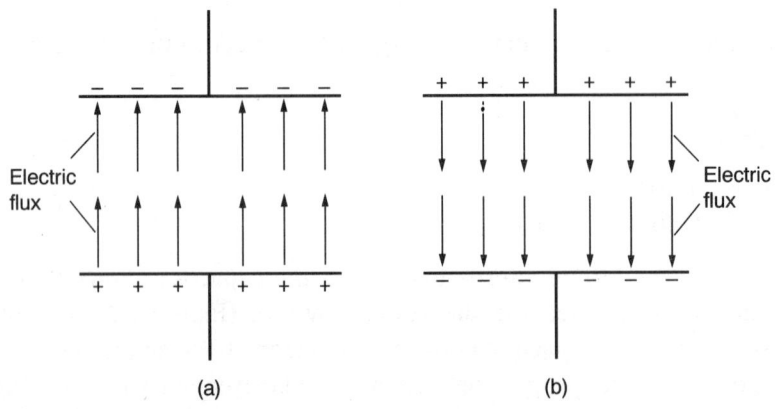

Fig. 2.24 (a) Charge distribution on positive half-cycles; and (b) charge distribution on negative half-cycles

Let us now see what happens when the supply voltage has a negative half-cycle. Negative charges on the bottom plate will attract positive charges to the top plate, and repel any negative charges. Thus, in the negative half-cycle of the supply voltage, negative charges on the lower plate induce positive charges on the upper plate.

This study of charge build-up shows why there is a phase difference between the supply voltage and the current – positive charges on one plate induce negative charges on the other plate, and vice versa. It also shows that, although charges appear to flow across the dielectric, they do not in reality – it is the electric flux that flows through the dielectric. If the supply voltage varies with time, the electric flux will also vary with time. Now, the units of electric flux are the same as charge, and so if we calculate the rate of change of flux with time, we will get units of coulomb per second, or amps, i.e.

$$\text{rate of change of flux} = \frac{d\psi}{dt}$$

From Gauss' law we know that each unit charge radiates a unit of electric flux, i.e. $\psi = Q$. So, the rate of change of flux is the same as the rate of flow of charges in the wires connected to the capacitor, i.e.

$$\text{rate of change of flux} = \frac{dQ}{dt}$$

$$= \text{capacitor current}$$

As the units of $d\psi/dt$ are the same as for current, we could regard $d\psi/dt$ as a current. In view of this it is known as displacement current because it displaces charges from the opposite plate of the capacitor.

So, this 'fields' model of a capacitor has explained why there is a phase shift between the supply voltage and the current. It has also introduced us to the idea of displacement current. This is very important when considering electromagnetic radiation, or radio waves, and it is essential that readers are happy with the concept. Figure 2.25 summarizes the mechanism by which current apparently flows through a capacitor.

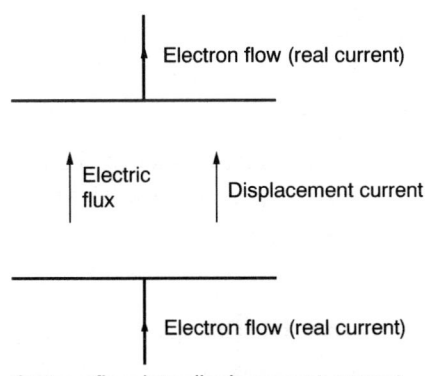

Fig. 2.25 Conversion of electron flow into displacement current

Example 2.15

A 22 µF capacitor is connected to a 4 V a.c. supply which has a frequency of 100 Hz. Determine the current taken from the supply. In addition, calculate the displacement current in the capacitor dielectric.

Solution

The frequency of the supply is 100 Hz, and so the angular frequency is

$$\omega = 2\pi f$$
$$= 2\pi \times 100$$
$$= 200\pi \text{ rad s}^{-1}$$

Now, the reactance of the capacitor is given by

$$X_C = \frac{1}{\omega C}$$

and so,

$$X_C = \frac{1}{200\pi \times 22 \times 10^{-6}}$$
$$= 72.34 \text{ ohm}$$

Thus the supply current is

$$i_s = \frac{V_s}{X_C}$$
$$= \frac{4}{72.34}$$
$$= 55.3 \text{ mA}$$

As the 'real' current changes to displacement current when it encounters the dielectric, we can write

$$\text{displacement current} = \frac{dQ}{dt}$$
$$= 55.3 \text{ mA}$$

Of course, we can never get an ideal dielectric. Thus there will also be some flow of charge across the capacitor, i.e. there will be some 'real' current. We will meet this again in Section 4.4.

Capacitance as resistance to flux

When considering the production of fields in a capacitor, it is sometimes useful to regard the capacitance as the resistance to the flow of electric flux. We have already seen, in Equation (2.35), that

$$Q = CV \qquad (2.61)$$

and so, by applying Gauss' law,

$$\psi = CV \qquad (2.62)$$

or,

$$V = \psi \frac{1}{C} \qquad (2.63)$$

Equation (2.63) relates the potential across the capacitor to the electric flux by way of the inverse of the capacitance. So, the higher the capacitance of a conductor system, the easier it is to produce electric flux. Thus we can regard the capacitance as a measure of the resistance to electric flux. (Although some readers may be wondering why this point is being stressed, all will become clear when we compare electrostatics, electromagnetism and electroconduction in Chapter 5.)

Example 2.16

A 10 μF capacitor has a potential of 100 V d.c. across its terminals. Determine the flux through the capacitor. If the capacitance increases to 50 μF, determine the new flux.

Solution

We have a 10 μF capacitor with 100 V across it. Thus the flux through the capacitor is (Equation (2.62)),

$$\psi = CV$$
$$= 10 \times 10^{-6} \times 100$$
$$= 1 \text{ mC}$$

The capacitance is now increased to 50 μF, and so the new flux is

$$\psi = CV$$
$$= 50 \times 10^{-6} \times 100$$
$$= 5 \text{ mC}$$

So, by increasing the capacitance we have increased the flux through the capacitor. These fluxes are, of course, equal to the charge stored in the capacitor.

Combinations of capacitors

In engineering we often have to make something new out of existing components. This is because manufacturers like to produce standard components to

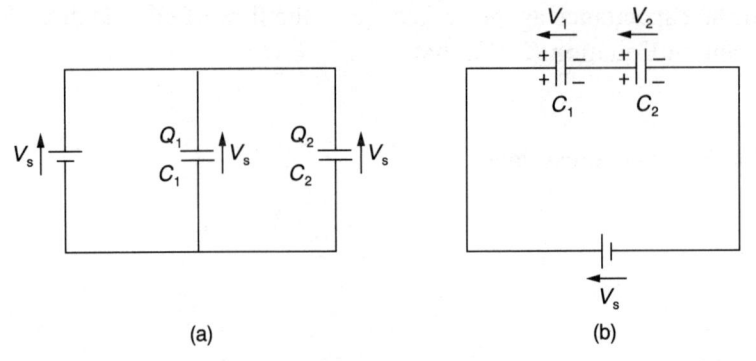

Fig. 2.26 (a) Parallel connection of capacitors; and (b) series connection of capacitors

keep costs down. When designing a circuit we often need a particular value of capacitor that is not available from any source. We could pick a value close to the one we require, and then physically alter the area of the capacitor to get the capacitance required by filing the top down. This requires the use of a capacitance meter, a file, a steady hand, and a great deal of patience! Although this can be done, it is not a very practical way to get a non-standard value of capacitance. Instead, we can produce non-standard capacitance values by combining standard values in parallel or series.

Figure 2.26(a) shows two capacitors, C_1 and C_2, in parallel. We require to find the equivalent capacitance of this arrangement. Let us connect a d.c. source, V_s, to the capacitors. Now, both capacitors will have the same voltage across them, but store different charge. Thus we can write

$$Q_1 = C_1 V_s \tag{2.64a}$$

and

$$Q_2 = C_2 V_s \tag{2.64b}$$

If we replace the two capacitors by a single equivalent one of value C_t, the charge on the new capacitor, Q_t, must be the same as on the parallel combination. Thus,

$$Q_t = C_t V_s \tag{2.65}$$

As Q_t must be the sum of the individual charges, in order to be equivalent, we can write

$$Q_t = Q_1 + Q_2$$

or,

$$C_t V_s = C_1 V_s + C_2 V_s$$

Thus,

$$C_t = C_1 + C_2 \tag{2.66}$$

So, we can increase capacitance by adding another capacitor in parallel with the original.

Let us now consider a series combination of capacitors – Fig. 2.26(b). As before, we will connect this combination to a d.c. source, and replace the capacitors with an equivalent one that will hold the same charge.

Now, let us assume that, when connected to the supply, a positive charge builds up on the left-hand plate of C_1. This charge induces an equal negative charge on the right-hand plate of C_1. The negative charge on this plate has to come from the left-hand plate of C_2, which leaves this plate positively charged. This positive charge on the left-hand plate induces a negative charge on the right-hand plate of C_2. As no charge leaves the circuit, the two capacitors store the same charge, but have different voltages across them. So, we can write

$$Q_1 = Q_2$$

or,

$$C_1 V_1 = C_2 V_2 \qquad (2.67)$$

Now, we are seeking to replace this series combination by a single capacitor of value C_t which must store a charge of Q_t when connected to a supply of V_s. Also, as the supply voltage must equal the individual voltage drops around the circuit, we can write,

$$V_t = V_1 + V_2$$

and so,

$$\frac{Q_t}{C_t} = \frac{Q_1}{C_1} + \frac{Q_2}{C_2}$$

The charge in the circuit is a constant given by

$$Q_t = Q_1 = Q_2$$

and so,

$$\frac{Q_1}{C_t} = \frac{Q_1}{C_1} + \frac{Q_1}{C_2}$$

which gives,

$$\frac{1}{C_t} = \frac{1}{C_1} + \frac{1}{C_2} \qquad (2.68)$$

So, we can decrease capacitance by adding another capacitor in series with the original.

Example 2.17

A 10 μF capacitor is connected in a circuit. What is the effect of placing a 1 μF capacitor in parallel with it? What happens if the 1 μF is connected in series with the original?

Solution

We have a 10 μF capacitor, and a 1 μF connected in parallel. So, the total capacitance is

$$C_t = 10 + 1$$
$$= 11 \, \mu F$$

which indicates that the capacitance has barely altered.

If we now connect the 1 μF capacitor in series, we get a new capacitance of

$$\frac{1}{C_t} = \frac{1}{10} + \frac{1}{1}$$
$$= 0.1 + 1$$
$$= 1.1$$

i.e.

$$C_t = 0.91 \, \mu F$$

So, the 1 μF capacitor dominates the 10 μF capacitor when it is in series with the larger capacitor.

2.10 SOME APPLICATIONS

Although we may not realize it, we use electrostatics every time we turn on a piece of equipment containing a cathode ray tube (CRT). The domestic television set uses a CRT, as do most computer monitors. Figure 2.27(a) is a photograph of a typical CRT. At the rear is a long narrow neck housing the electron gun. This device accelerates electrons to speeds approaching that of light, and Fig. 2.27(b) shows a schematic of the structure in greater detail. At the very rear of the electron gun is a heater that causes the cathode to emit electrons. (These electrons are thermally excited because the cathode reaches temperatures in excess of 1800 °C. This is thermionic emission.)

Of course, if the CRT contains air, the electrons will lose momentum, and may fail to reach the phosphor screen at the front. This will result in loss of picture; not very desirable. Thus it is important that the CRT has a vacuum inside. This is usually done by burning magnesium in the tube after it has been sealed.

Now for some simple calculations. Let us assume that the anode is grounded, and that the cathode is at a negative voltage of V volt. Let us also assume that

2.10 Some applications

(a)

(b)

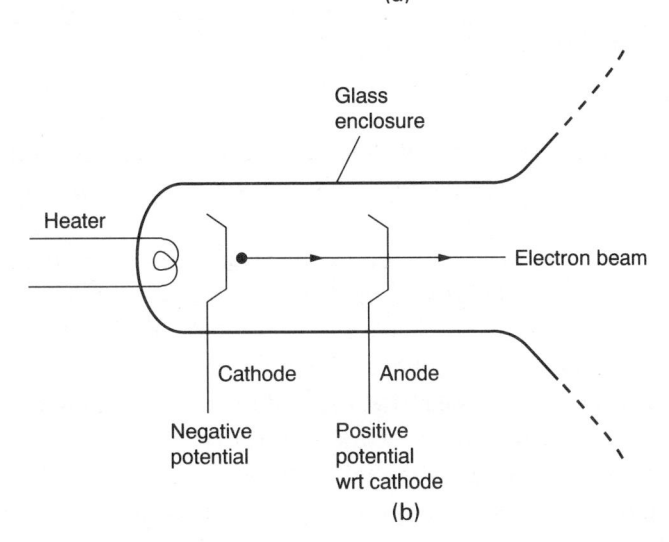

Fig. 2.27 (a) A typical cathode ray tube (CRT); and (b) basic structure of the electron gun

the distance between the anode and cathode is x metre. When the cathode produces electrons by thermionic emission, they are accelerated by the $\boldsymbol{E}$ field. So, the force acting on one electron is

$$\boldsymbol{F} = q\boldsymbol{E} \tag{2.69}$$

This must be equal to the mass of the electron, m, times the acceleration due to the field, $\boldsymbol{a}$. Thus,

$$\boldsymbol{F} = q\boldsymbol{E} = m\boldsymbol{a}$$

giving

$$a = \frac{q}{m}E \tag{2.70}$$

If we assume that the electrons are initially at rest, we can find their final velocity using

$$v^2 = 2as$$

and so,

$$v^2 = 2\frac{q}{m}Ex$$

$$= 2\frac{q}{m}\frac{V}{x}x$$

$$= 2\frac{q}{m}V$$

We can rearrange this equation to give

$$\tfrac{1}{2}mv^2 = qV \tag{2.71}$$

Equation (2.71) is simply a form of the conservation of energy – the left-hand side is the increase in kinetic energy, while the right-hand side is the electron energy in electron-volts.

As an example, let us take a cathode voltage of -20 kV. As the mass of an electron is 9.1×10^{-31} kg, this gives a final velocity of 8.4×10^7 m s^{-1}. This is the velocity of the electrons as they leave the neck of the CRT. They then travel across the wider part of the tube, where they enter a magnetic field that deflects their path. This is briefly considered at the end of the next chapter.

We can also find electrostatics at work in industry. The air in a factory often contains oil and dust particles. Although some form of vacuum cleaner can be used, the maintenance costs are quite high. An alternative is to ionize the particles, and use electrostatics to attract them to charged metal plates where they can be removed. Such a device is the electrostatic precipitator, and Fig. 2.28 shows the schematic diagram of a typical example.

As can be seen from the figure, the precipitator basically consists of a wire placed between two metal plates. In operation, the wire is maintained at a large negative potential, typically -50 kV, with respect to the plates. Under these conditions, the E field close to the wire is large enough to ionize the air. The negative field repels the freed electrons, while the wire attracts the positive charges. So, electrons are accelerated toward the outer plates.

Now, if there are dust particles between the plates, the free electrons will attach themselves to the dust, so making them negatively charged. Thus the

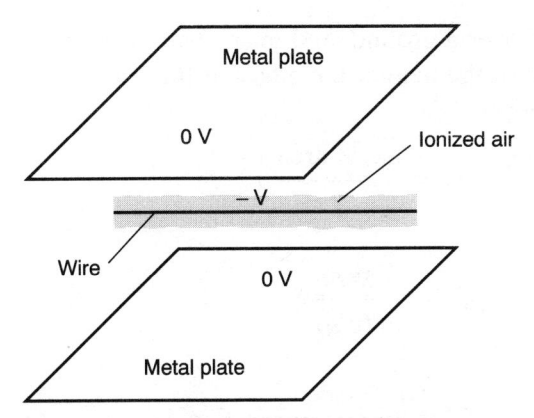

Fig. 2.28 Schematic diagram of a typical electrostatic precipitator

positively charged plates attract the negatively charged particles, so removing them from the atmosphere.

The precise analysis of this situation is complicated because the plates are not coaxial to the wire. In this case we must plot the electrostatic field strength between the plates to find out what happens to the dust particles. Under such circumstances it is probably easier to build a prototype and experiment.

2.11 SUMMARY

We started this chapter by examining a fundamental law relating to the force between point charges – Coulomb's law. We then went on to develop the ideas of electric flux, electric field, and potential. Again, we were only concerned with point charges. The relevant formulae are summarized here:

$$F = \frac{q_1 q_2}{4\pi \epsilon r^2} r \quad (2.72)$$

$$D = \frac{q_1}{4\pi r^2} r \quad (2.73)$$

$$E = \frac{q_1}{4\pi \epsilon r^2} r \quad (2.74)$$

$$V = \frac{q_1}{4\pi \epsilon r} \quad (2.75)$$

We then considered certain charge distributions: line charges; surface charges; and volume charges. In all of these cases we were able to apply Gauss' law to simplify the analysis.

We next examined various types of capacitor: parallel plate; coaxial cable;

twin feeder; wire over a ground; and microstrip lines. In all cases we chose to ignore the effects of the field at the edges of the conductors. The capacitances are reproduced here:

parallel plate $$C = \frac{\epsilon_0 \epsilon_r \times \text{area}}{d} \text{ F} \qquad (2.76)$$

coaxial cable $$C' = \frac{2\pi\epsilon_0\epsilon_r}{\ln(b/a)} \text{ F m}^{-1} \qquad (2.77)$$

twin feeder $$C' = \frac{\pi\epsilon}{\ln((d-a)/a)} \text{ F m}^{-1} \qquad (2.78)$$

wire over ground $$C' = \frac{2\pi\epsilon_0}{\ln((d/2-a)/a)} \text{ F m}^{-1} \qquad (2.79)$$

microstrip $$C' = \frac{\epsilon_0\epsilon_r w}{h} \text{ F m}^{-1} \qquad (2.80)$$

or, $$C' = \frac{2\pi\epsilon_0\epsilon_r}{\ln(h/w)} \text{ F m}^{-1} \qquad (2.81)$$

We also encountered the fundamental formula:

$$Q = CV \qquad (2.82)$$

We then went on to consider the storage of energy by a capacitor. We found that the energy can be regarded as either stored on the capacitor plates, or stored in the electrostatic field between the plates. This is an important concept to grasp as it shows the equivalence between a field theory approach, and the more familiar 'circuits' approach. The stored energy is given by

$$\text{energy} = \tfrac{1}{2}CV^2 \text{ J} \qquad (2.83)$$

or,

$$\text{energy} = \tfrac{1}{2}DE \text{ J m}^{-2} \qquad (2.84)$$

The question of how the current 'flowed' through an ideal capacitor was then examined. This introduced us to the idea of displacement current, and showed us that the 'real' current converted to 'displacement' current and back again as it crossed the dielectric. This is a very important point, which we cannot explain by the 'circuits' approach. Indeed, we found that we can regard capacitance as the 'resistance' to the flow of flux. This acts as a link between the fields and circuits approaches.

We also saw that the reactance of a capacitor is given by

$$X_C = \frac{1}{2\pi f C} \tag{2.85}$$

with the capacitor current leading the supply voltage by 90°.

Next we examined parallel and series combinations of capacitors. We saw that capacitance can be increased by adding another capacitor in parallel with the original, and decreased by adding series capacitance.

We concluded this chapter with a brief examination of two applications of electrostatics: electron acceleration in a cathode ray tube; and electrostatic precipitation. In the first example we saw that electrons can be accelerated to very high velocities by a potential difference of, typically, 20 kV. Electrostatic precipitators also use potentials of this order to attract dust and smoke particles to metal plates, so cleaning the air.

3 Electromagnetic fields

When we considered electrostatics in the previous chapter, we started with the force between isolated point charges. This introduced us to the ideas of flux density and electric field strength. Now that we are considering magnetism, we can also start at the same point – isolated north or south monopoles – and Section 3.1 develops some basic ideas based around this concept. However, no-one has yet found isolated magnetic poles and so we will quickly encounter the magnetic field generated by a current-carrying elemental wire in Section 3.2 – hence the term electromagnetism. Once we have grasped this idea, we will leave magnetic monopoles behind. (Of course, if someone does find magnetic monopoles, we will have to rewrite all the textbooks – this one included!)

3.1 SOME FUNDAMENTAL IDEAS

At about the same time that Coulomb was examining the force between isolated charges, he was also experimenting with magnetism (1785). In common with electrostatics, he found that the force between two magnetic poles decreases as the inverse of the square of the distance separating them, i.e.

$$F = \frac{p_1 p_2}{kr^2} r \qquad (3.1)$$

where F is the vector force between the two poles (N)
p_1 and p_2 are the strengths of the magnetic poles (Wb)
k is a constant of proportionality
r is the distance between the two poles (m),
and r is the unit vector acting in the direction of the line joining the two charges.

This is the exact parallel of Coulomb's law as applied to isolated point charges. The force is repulsive if the poles are alike, and attractive if the poles are dissimilar (Fig. 3.1).

We can extract a factor of 4π from the constant k to give

$$F = \frac{p_1 p_2}{4\pi\mu r^2} r \qquad (3.2)$$

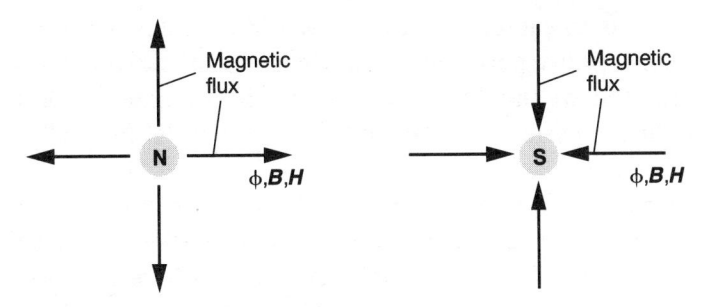

Fig. 3.1 Two separate magnetic monopoles in free space

where μ is the permeability – a material property. (If we use the SI system of units, the force is in newton if the pole strengths are in weber (named after Wilhelm Eduard Weber, 1804–1891, the German physicist noted for his study of terrestrial magnetism), μ is in H m^{-1}, and r is in metre. The reason for the choice of units for μ will become clear when we consider inductance in Section 3.11.)

It is now a simple matter to introduce the idea of magnetic field strength in the same way that we introduced electric field strength. Thus, the force on a magnetic pole of strength p_2 is

$$\boldsymbol{F} = p_2 \boldsymbol{H} \tag{3.3}$$

where $\boldsymbol{H}$ is the magnetic field strength given by

$$\boldsymbol{H} = \frac{p_1}{4\pi\mu r^2}\boldsymbol{r} \tag{3.4}$$

with units of newton per weber.

If we adapt Gauss' law to magnetostatics, we can say that the magnetic flux emitted by a pole is equal to the strength of the pole. Thus we can define the magnetic flux density as

$$\boldsymbol{B} = \frac{p_1}{4\pi r^2}\boldsymbol{r} \tag{3.5}$$

with units of Wb m^{-2}.

We can combine Equations (3.4) and (3.5) to give

$$\boldsymbol{B} = \mu \boldsymbol{H} \tag{3.6}$$

In common with electrostatics, the value of the constant of proportionality (in this case the permeability) is dependent on the material. When working with magnetism, it is common practice to work with the permeability relative to free space. Thus,

$$\mu_r = \frac{\mu}{\mu_0} \tag{3.7}$$

where μ_0 is the permeability of free space with value $4\pi \times 10^{-7}\,\text{H m}^{-1}$. Unfortunately the relative permeability of a magnetic material varies according to the flux density. So we shouldn't really refer to it as a material constant. (We will meet the relative permeability again when we consider magnetic materials in Chapter 7.)

So, we have developed a model in which magnetic flux emanates from an isolated pole (assumed to be a point source) in a radial direction. This model is identical to that adopted for isolated point charges in Chapter 2. However, we must use caution from this point onwards. This is because no-one has yet found isolated magnetic monopoles. Instead we must adapt our model to current-carrying wires.

Example 3.1

A single 10 μWb magnetic monopole is situated in air. Calculate the magnetic field strength at a distance of 0.5 m from the monopole. In addition, find the flux density and the force on an identical monopole at the same distance.

Solution

We have a single monopole of strength 10 μWb situated in air. Now, the magnetic field strength is (Equation (3.4))

$$H = \frac{p_1}{4\pi \mu r^2}\, r$$

and so,

$$H = \frac{10 \times 10^{-6}}{4\pi \times 4\pi \times 10^{-7} \times 0.5^2}\, r$$

$$= 2.53\, r\ \text{N Wb}^{-1} \quad \text{in a radial direction}$$

From Equation (3.6) we have

$$B = \mu H$$

and so,

$$B = 4\pi \times 10^{-7} \times 2.53\, r$$
$$= 3.2 \times 10^{-6}\, r$$
$$= 3.2\, r\, \mu\text{Wb m}^{-2} \quad \text{in a radial direction}$$

If we place an identical monopole at 0.5 m from the original, the force on this monopole is

$$F = p_2 H$$
$$= 10 \times 10^{-6} \times 2.53\, r$$
$$= 25.3\, r\ \mu\text{N} \quad (\text{repulsive})$$

So, even if we have low-strength monopoles, the field strength can be quite high (2.53 N Wb^{-1}). However, even with such a high value of H, the flux density is low (3.2 μWb m^{-2}). These values are quite typical when considering magnetism – the magnetic field strength can be quite high, but the flux density will be low. (Particle accelerators and fusion reactors use exceptionally high flux densities, of values greater than 1 Wb m^{-2}.)

3.2 SOME ELEMENTARY CONVENTIONS USED IN ELECTROMAGNETISM

Having established that the inverse square law applies to isolated magnetic monopoles, we will now examine the magnetic field produced by a current-carrying conductor. (Although readers may not be very familiar with this effect, everyone has come across it. Any piece of rotating electrical equipment – power tools, alternators, starter motors, etc. – relies on the magnetic field produced by a current-carrying wire. We will consider a simple motor/generator in Chapter 7.)

In 1819, Hans Christian Oersted (a Danish physicist) demonstrated that a magnetic field surrounds a current-carrying wire. This was a very important discovery because it unified the separate sciences of electricity and magnetism into one science – electromagnetism. (Indeed, this was the first indication that two different forces of Nature could be unified. The search is now on for a Grand Unified Theory that would explain life, the Universe, and everything!)

Oersted plotted the field surrounding a current-carrying wire using a compass. As Fig. 3.2 shows, the magnetic field is coaxial to the wire. The direction of the field depends on whether the current flows up, as in Fig. 3.2(a), or down the wire, as in Fig. 3.2(b).

We now come across some conventions:

(1) **A cross denotes current flowing into the page. A dot denotes current flowing out of the page.**

Figure 3.2(c) shows these two conventions.

Readers who play darts might like to imagine a dart thrown into the page. As the dart travels away from us, we see the cross of the feathers. Thus a cross denotes current travelling away from us, down the wire. If the dart is travelling towards us, we see the point of the dart first – until it hits us! Hence a dot corresponds to current travelling towards us, up the wire.

(2) **The right-hand corkscrew rule gives the direction of the flux.**

Figure 3.2(c) shows this rule.

Most of us are familiar with the action of a corkscrew. We can use this action to determine the direction of the magnetic field. If we have current

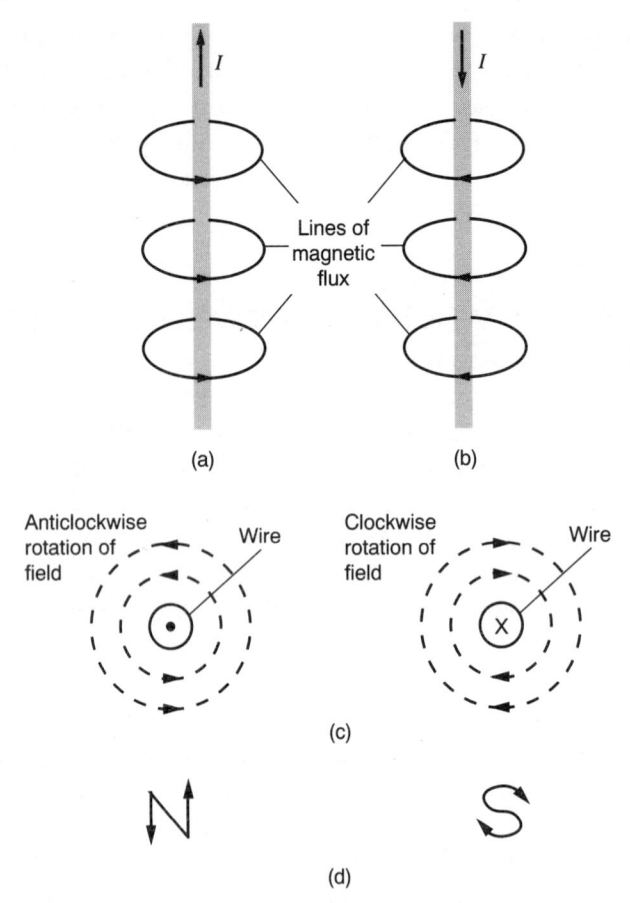

Fig. 3.2 (a) Magnetic field produced by upward flowing current; (b) magnetic field produced by downward flowing current; (c) dot and cross conventions of current flow; and (d) north and south pole conventions

flowing away from us into the page, a corkscrew will have to turn in a clockwise direction to follow the current. Thus the field acts in a clockwise direction. If the current is flowing towards us out of the page, the corkscrew acts in an anticlockwise direction. Thus the field acts in an anticlockwise direction.

(3) **Clockwise rotation of the field denotes a south pole. Anticlockwise rotation of the field gives a north pole.**

We can easily apply this rule by drawing little arrows on the letter N (for north) and S (for south) as Fig. 3.2(d) shows.

So, Oersted showed that a current-carrying wire generates a coaxial magnetic field. The field lines that we have drawn in Fig. 3.2 are lines of magnetic flux. It is important to note that these flux lines act in a completely different direction to those we met in the previous section. This is a very important point to grasp

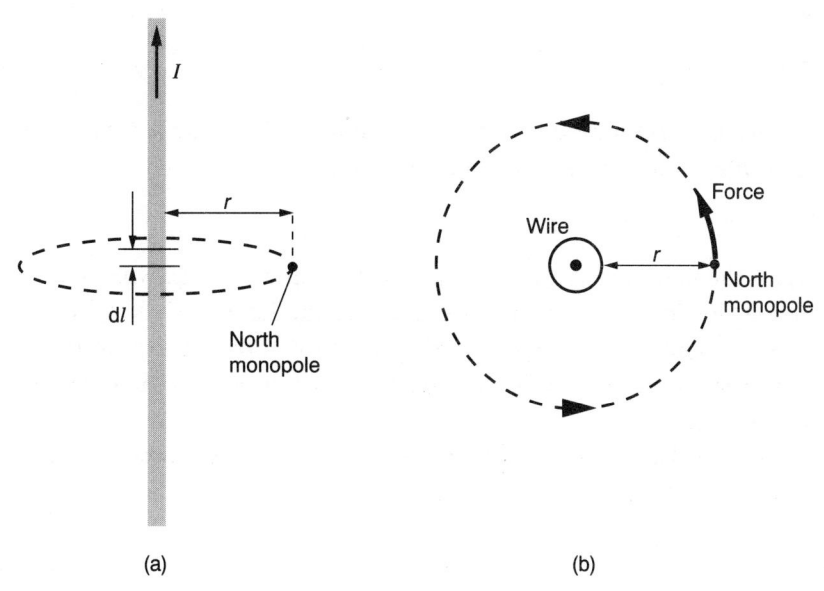

Fig. 3.3 (a) Magnetic field produced by a current-carrying wire; and (b) plan view of wire/magnetic field

– when we consider the field surrounding a current-carrying wire, we cannot use the simple monopole model.

In the next section, we will examine the magnetic field produced by a simple current element – the Biot–Savart law.

3.3 THE BIOT–SAVART LAW

Let us consider the current-carrying conductor shown in Fig. 3.3. Oersted showed that this conductor will generate a magnetic field that is coaxial to the wire. So, if we place an imaginary unit north pole at a point P, distance r from a small elemental section of the wire of length dl, the wire will experience a force that will tend to push it to the left. (The plan view in Fig. 3.3(b) shows why this is so.) In addition, there will be an equal and opposite force on the north pole due to the field surrounding the wire.

Let us try to find the magnetic field strength, δH_1, at the north pole, due to the current element formed by I and dl. As we have just discussed, the current element produces a force on the north pole, and the north pole will produce an equal and opposite force on the current element. If we can find these two forces, and then equate them, we should get an expression for the magnetic field strength generated by the wire.

Let us initially consider the field at dl due to the imaginary north pole of strength p_N. As this north pole is a point source, it emits magnetic flux in a radial direction. Thus we can write the flux density as

70 Electromagnetic fields

$$\boldsymbol{B}_N = \frac{p_N}{4\pi r^2} \boldsymbol{r} \tag{3.8}$$

Direct experimental measurement shows that the force on a current-carrying conductor placed in a magnetic field is given by

$$F = BIl \tag{3.9}$$

where B is the flux density of the magnetic field in which the wire is placed, I is the current flowing through the wire, and l is the length of the wire. (We can intuitively reason that this equation is correct by noting that powerful electric motors require a large electric current, and contain a large amount of wire – they are very heavy!)

By combining Equations (3.8) and (3.9), we find that the force on the element dl due to the field emitted by the north pole is

$$dF = \frac{p_N}{4\pi r^2} I \, dl \tag{3.10}$$

Let us now turn our attention to the force on the north pole produced by the current element. The current element formed by the current I and length dl produces a magnetic field strength of dH_1 at the north pole. By applying Equations (3.3) and (3.4), the force on the north pole due to the current element is

$$dF = p_N \, dH_1 \tag{3.11}$$

By equating Equations (3.10) and (3.11) we get

$$\frac{p_N}{4\pi r^2} I \, dl = p_N \, dH_1$$

and so,

$$d\boldsymbol{H}_1 = \frac{I \, dl}{4\pi r^2} \boldsymbol{\phi} \tag{3.12}$$

with direction into the page. ($\boldsymbol{\phi}$ is the unit vector acting into the page.) This is the magnetic field strength due to the small current element formed by I and dl. To find the total field produced by the wire, we should integrate this equation with respect to length. However, Equation (3.12) only gives the field at a point directly opposite the current element we are considering. To find the field we require, we must do some resolving of components.

Figure 3.4 shows the situation we now seek to analyse. If we draw a line from the point P to the current element, we find that it makes an angle to the current element of θ. Now, if the wire is infinitely long, θ will vary from 0 to π. We can intuitively reason that at the two extremes of the wire, $+\infty$ and $-\infty$, the force on a north pole placed at the point P will be zero. So, if we modify Equation (3.12) by a factor $\sin \theta$ we will get

3.4 Electromagnetic flux, flux density and field strength

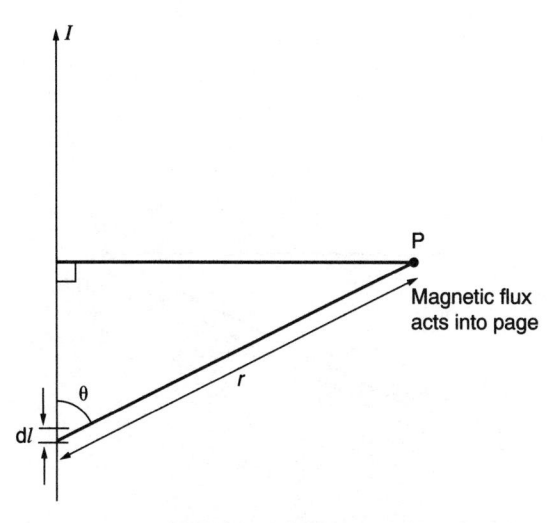

Fig. 3.4 Construction for finding the magnetic field of an infinitely long current-carrying wire

$$dH = \frac{I\,\delta l}{4\pi r^2}\sin\theta \tag{3.13}$$

with direction into the page. (We can check this equation by noting that $\sin 0$ and $\sin \pi$ are both zero. Thus the force on the north pole generated by current elements at the extremes of the wire, at $+\infty$ and $-\infty$, is zero. The use of $\sin\theta$ in Equation (3.13) is deliberate: if we use vector algebra, the H field is proportional to the cross product of I and the vector δl drawn normal to the wire. The cross product introduces the factor $\sin\theta$.)

Although Equation (3.13) is known as the Biot–Savart law, we should really credit it to Ampère (1820) who originally did experiments on current-carrying conductors. (Biot and Savart were colleagues of Ampère.) In order for this equation to be of any practical use, we must integrate the current elements over the length of our conducting wire. This we will do in Section 3.6.

We will now re-examine the idea of magnetic flux, magnetic flux density, and magnetic field strength as produced by a current element.

3.4 ELECTROMAGNETIC FLUX, FLUX DENSITY AND FIELD STRENGTH

When we considered electrostatics in the previous chapter, we made use of Gauss' law to state that the electric flux emitted by a positive charge was equal to the enclosed charge. Now that we are considering electromagnetic fields, we can adapt Gauss' law to state that the magnetic flux generated by a current element is equal to the product of the current and element length. Thus,

$$p = I\,dl \text{ weber (Wb)} \tag{3.14}$$

72 Electromagnetic fields

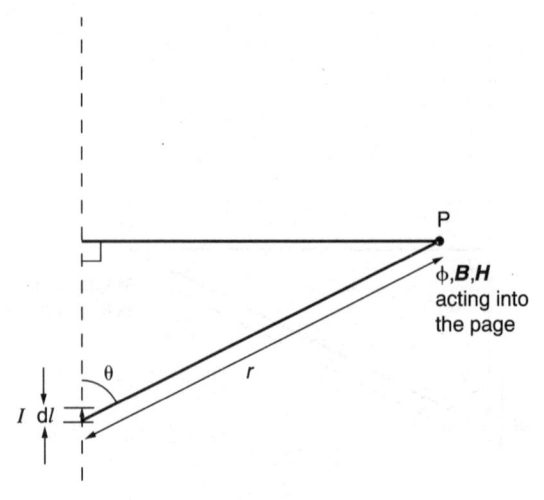

Fig. 3.5 Magnetic field due to an isolated current element

As we have just seen, if we consider the isolated current element, $I\,dl$, of Fig. 3.5, we have a fractional magnetic field strength, dH, of

$$dH = \frac{I\,dl}{4\pi r^2} \sin\theta \text{ A m}^{-1} \tag{3.15}$$

acting into the page. If we introduce a north pole at the point P, Equation (3.11) shows that it will experience a force of

$$dF = \frac{I\,dl}{4\pi r^2} \sin\theta\, p_N \tag{3.16}$$

So, the force on a magnetic pole is directly dependent on the magnetic field strength produced by the wire. This agrees with our model of electrostatics.

As $B = \mu H$, the fractional magnetic flux density is given by

$$dB = \mu \frac{I\,dl}{4\pi r^2} \sin\theta \tag{3.17}$$

We should note that if we introduce a current element of strength p_i at the point P, the force on this element will be given by

$$dF = \mu \frac{I\,dl}{4\pi r^2} \sin\theta\, p_i$$

or,

$$dF = dB\, p_i \tag{3.18}$$

Thus the force on a current element is directly dependent on the magnetic flux density produced by the wire, and not the magnetic field strength. We must exercise great care here as this is often a source of confusion.

We have now completed our initial study of magnetostatics and electromagnetism. In the next section we will compare the fundamental equations of electrostatics, magnetostatics, and electromagnetism.

3.5 COMMENT

We can now compare our models of electrostatics, magnetostatics and electromagnetism. Table 3.1 lists the fundamental formulae we have met.

Table 3.1 Comparison of fundamental formulae for electrostatics, magnetostatics, and electromagnetism in free space

	Force	*Field strength*	*Flux density*
Electrostatics	$F = \dfrac{q_1 q_2}{4\pi\epsilon_0 r^2} = q_2 E$	$E = \dfrac{q_1}{4\pi\epsilon_0 r^2}$	$D = \dfrac{q_1}{4\pi r^2}$
Magnetostatics	$F = \dfrac{p_1 p_2}{4\pi\mu_0 r^2} = p_2 H$	$H = \dfrac{p_1}{4\pi\mu_0 r^2}$	$B = \dfrac{p_1}{4\pi r^2}$
Electromagnetism	$dF = \dfrac{I\,dl}{4\pi r^2} p_2 \sin\theta = p_2\,dH$ (force on a magnetic pole) $dF = \mu_0 \dfrac{I\,dl}{4\pi r^2} p_i \sin\theta = p_i\,dB$ (force on a current element)	$dH = \dfrac{I\,dl}{4\pi r^2} \sin\theta$	$dB = \mu_0 \dfrac{I\,dl}{4\pi r^2} \sin\theta$

A glance at Table 3.1 shows a lot of similarity between electrostatics and magnetostatics. We should expect this because we developed our magnetostatic model along the lines of our electrostatic model. However, there are some obvious differences between electrostatics and electromagnetism: in electrostatics the flux density is independent of any change in the surrounding material, whereas in electromagnetism, the flux density is directly proportional to a material property (the permeability). As we will see in Chapter 7, this is due to the ability of certain materials to increase the flux density produced by a coil of wire.

We can also make one further observation: in electrostatics, the force on an isolated point charge is dependent on the field strength. Similarly, in electromagnetism, the force on an isolated magnetic pole is dependent on the field strength, but the force on a current element depends on the flux density. This can be a source of confusion, and we must exercise great care here. (This is because the science of magnetostatics developed in parallel with electrostatics, with isolated poles and charges being assumed in both cases. As we have seen, this means that the fundamental formulae for electrostatics and magnetostatics are very similar. With Oersted's discovery of electromagnetism, it was soon realized that it was the magnetic flux density, and not the field strength

74 Electromagnetic fields

produced by a current element, that determines the force on a current-carrying wire. So, we now have electromagnetic terms that do not match their electrostatic counterparts. The situation might have been different if Oersted had discovered electromagnetism before magnetostatics had been formalized!)

In the following sections we will apply the Biot–Savart law to a variety of current-carrying conductors. We begin by studying the field produced by a long, current-carrying conductor. This will introduce us to Ampère's circuital law. (Although Ampère's circuital law came before the Biot–Savart law, the way our electromagnetic model has developed dictates this approach.)

3.6 MAGNETIC FIELD STRENGTH AND AMPERE'S CIRCUITAL LAW

Let us consider the current-carrying wire shown in Fig. 3.6. We want to find the magnetic field strength at a point P, distance R from the wire. To find $\boldsymbol{H}$ at this point, we will determine the field strength due to a small current element, and then integrate the result over the length of the wire.

By applying the Biot–Savart law, we get

$$dH = \frac{I\,dz}{4\pi r^2} \sin\theta \text{ A m}^{-1} \tag{3.19}$$

acting into the page. Now, to find the total field strength, we need to integrate Equation (3.19) with respect to length. Unfortunately, as we move along the wire, the distance r and the angle θ will vary. So, we need to do some substitution and manipulation before we can do any integration.

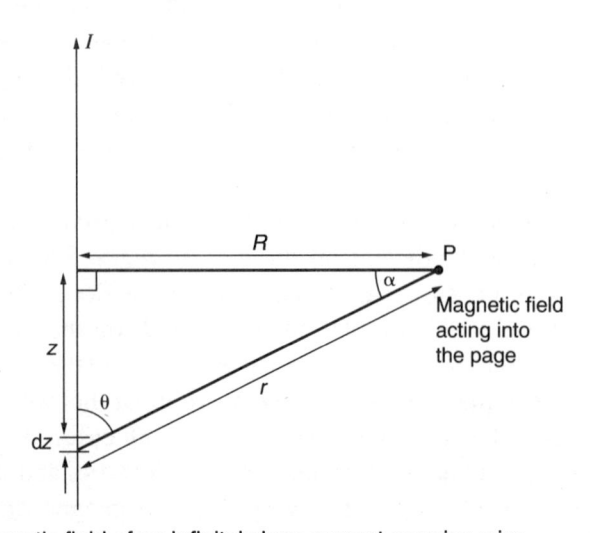

Fig. 3.6 The magnetic field of an infinitely long current-carrying wire

Instead of working with the angle θ, we can simplify the integration if we use the angle α instead. So, with reference to Fig. 3.6, we can see that $z = R \tan \alpha$, and so $dz = R\, d\alpha/\cos^2 \alpha$. As $\sin \theta = R/r = \cos \alpha$, we get $r = R/\cos \alpha$. Thus Equation (3.19) becomes

$$dH = \frac{I}{4\pi} \frac{R\, d\alpha}{\cos^2 \alpha} \frac{\cos^2 \alpha}{R^2} \cos \alpha$$

Now, as we move from $-\infty$ to $+\infty$ the angle α varies from $-\pi/2$ to $+\pi/2$. So,

$$H = \frac{I}{4\pi R} \int_{-\pi/2}^{+\pi/2} \cos \alpha \, d\alpha$$

$$= \frac{I}{4\pi R} \left| \sin \alpha \right|_{-\pi/2}^{+\pi/2}$$

$$= \frac{I}{4\pi R} (1 + 1)$$

and so,

$$H = \frac{I}{2\pi R} \quad \text{into the page.} \tag{3.20}$$

Let us take a moment to examine this equation in some detail. It shows that the magnetic field strength is proportional to the current in the wire. We should expect this because common sense tells us that a large magnetic field implies a large current. However, H is also inversely dependent on the term $2\pi R$. This is simply the circumference of a circle of radius R with the wire at the centre. We should also note that the field is independent of where exactly we are along the wire – provided we maintain a constant distance from the wire, the field is a constant. We can intuitively reason that these observations are correct because, as we have already seen, the magnetic field is coaxial with the wire.

In order to formalize this, we can write

$$I = \oint H\, dl \tag{3.21}$$

where $\oint$ denotes line integral around a closed loop. Equation (3.21) is Ampère's circuital law or, more simply, Ampère's law. (André Marie Ampère (1775–1836) was a French physicist who formulated this law.)

Example 3.2

A current of I A flows through a long straight wire, of radius a, situated in air. The current density in the wire is constant across the cross-section of the wire. Plot the variation of H with radius both inside and outside the wire.

Solution

We want to find the variation in magnetic field strength as a function of radius both inside and outside the wire. Now, the current density in the wire is independent of radius, i.e. $I/\pi r^2$ is a constant for $r < a$. So, the current enclosed by a circular path of radius r is

$$I' = \frac{I}{\pi a^2} \pi r^2 \quad r < a$$

Ampère's law, Equation (3.21), gives

$$I = \oint H \, dl$$

and so $I' = H_r \, 2\pi r$. Hence,

$$H_r = \frac{I'}{2\pi r}$$

$$= \frac{I}{\pi a^2} \pi r^2 \cdot \frac{1}{2\pi r}$$

$$= \frac{Ir}{2\pi a^2} \quad \text{for } r < a$$

i.e. the magnetic field is directly proportional to radius inside the wire.

Let us now consider a circular path outside the wire. This path encloses all the current, and so

$$I = H_r \, 2\pi r$$

Hence,

$$H_r = \frac{I}{2\pi r} \quad \text{for } r > a$$

i.e. the magnetic field is inversely proportional to radius outside the wire.

At the surface of the wire, these two values should be the same. So, as a check we can put $r = a$ to give

$$H_r = \frac{Ia}{2\pi a^2} = \frac{I}{2\pi a}$$

which is identical to the field outside the wire.

Figure 3.7 shows the variation of H with radius. If the wire is made of non-ferrous material, such as copper, the flux density will also follow the same variation.

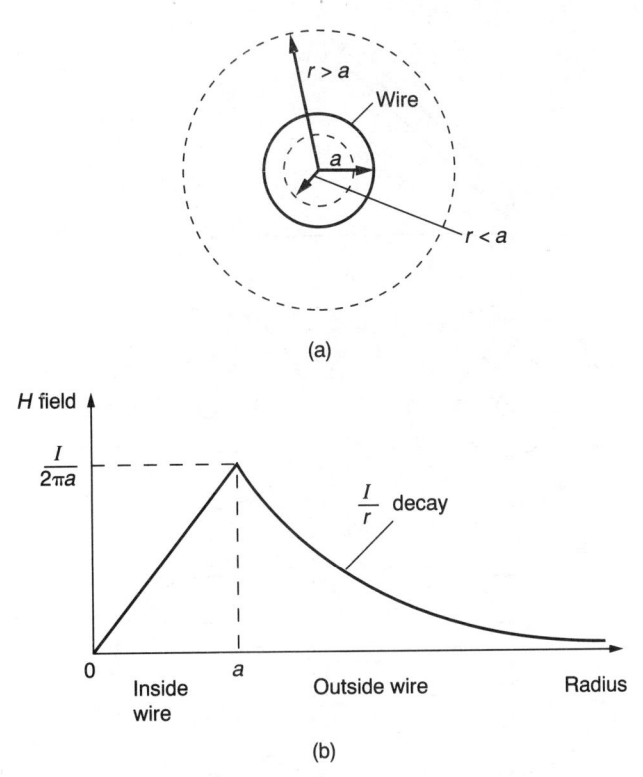

Fig. 3.7 (a) Circular paths inside ($r < a$) and outside ($r > a$) a wire; and (b) variation of H with radius

3.7 THE FORCE BETWEEN CURRENT-CARRYING WIRES – THE DEFINITION OF THE AMPERE

We can now define the ampere. Readers may think that this is not worth considering since the ampere is simply a measure of current set down by international treaty. After all, we do not often have to concern ourselves with the definition of the metre. However, as we will soon see, the definition of the ampere introduces us to the force between two current-carrying wires, and that is of some practical benefit.

Figure 3.8 shows the situation we are to analyse. Two current-carrying wires run parallel to each other, separated by a distance r. These wires each carry a current of I amperes. As we have already seen, current-carrying wires produce magnetic fields. As each wire carries the same current, the magnetic field produced by the left-hand conductor will exactly balance the field produced by the right-hand conductor at the point mid-way between the two conductors. Thus the field at this point will be zero, resulting in the field distribution of Fig. 3.8(c). The weakening of the field between the two wires shows that they attract each other.

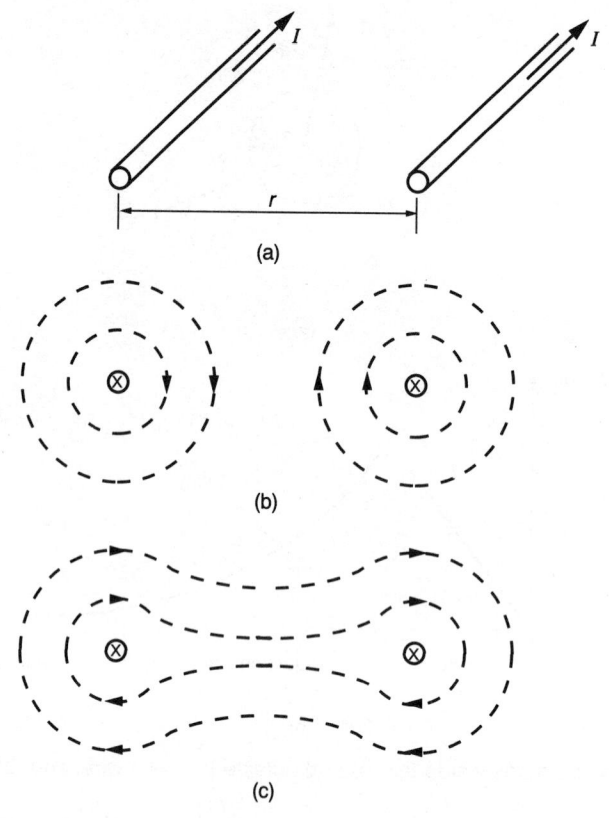

Fig. 3.8 (a) Two parallel current-carrying wires; (b) magnetic field produced by each wire; and (c) resultant field distribution

Now, in Section 3.4 we met the force on an isolated north pole due to a current-carrying element. In this example, we do not have an isolated pole; instead we must consider a current element in the right-hand wire.

To find the force on the right-hand conductor, we need to find the magnetic flux density produced by the left-hand conductor. By applying Ampere's law, Equation (3.21), we can write

$$I = \oint H \, dl$$

and so $I = H_1 \cdot 2\pi r$ or,

$$H_1 = \frac{I}{2\pi r} \tag{3.22}$$

at the surface of the right-hand conductor. As $B = \mu_0 H$, the flux density at the right-hand conductor is

$$B_1 = \frac{\mu_0 I}{2\pi r} \tag{3.23}$$

Now, the force on a current element in the right-hand wire will be (Equation (3.9))

$$dF = B_1 I\, dl$$
$$= \frac{\mu_0 I}{2\pi r} I\, dl \qquad (3.24)$$

To find the total force on the right-hand conductor, we should integrate this equation with respect to length. However, we can calculate the force per unit length by dividing both sides by dl. Thus,

$$\frac{dF}{dl} = \frac{\mu_0 I^2}{2\pi r}\, \text{N m}^{-1} \qquad (3.25)$$

If the separation between the two wires is 1 metre, the force between the two conductors is 2×10^{-7} N m^{-1}, and we use $4\pi \times 10^{-7}$ for μ_0, we get

$$\frac{dF}{dl} = 2 \times 10^{-7} = \frac{4\pi \times 10^{-7} \times I^2}{2\pi \times 1}$$

and so,

$I = 1$ ampere

At this point we should ask why a force of 2×10^{-7} N m^{-1} was chosen. The answer lies with the magnitudes of μ_0 and ϵ_0. The study of electromagnetic waves in space shows that μ_0 and ϵ_0 are related to the speed of light in a vacuum by

$$c = \frac{1}{\sqrt{\mu_0 \epsilon_0}} \qquad (3.26)$$

As the speed of light in a vacuum is 3×10^8 m s^{-1}, this fixes the relative values of μ_0 and ϵ_0, and so the force between the wires must be 2×10^{-7} N m^{-1}. (Of course, if we changed our system of units, the values of μ_0 and ϵ_0 would change. However, the SI system of units is in use today, and this gives us our values of μ_0 and ϵ_0.)

3.8 THE MAGNETIC FIELD OF A CIRCULAR CURRENT ELEMENT

In electrical engineering we often come across wound components – transformers and coils. Section 7.4 deals with transformers in detail. Here we concern ourselves with the field produced by a circular piece of wire carrying a current. This will help us when we come to consider coils and solenoids in the next section.

80 Electromagnetic fields

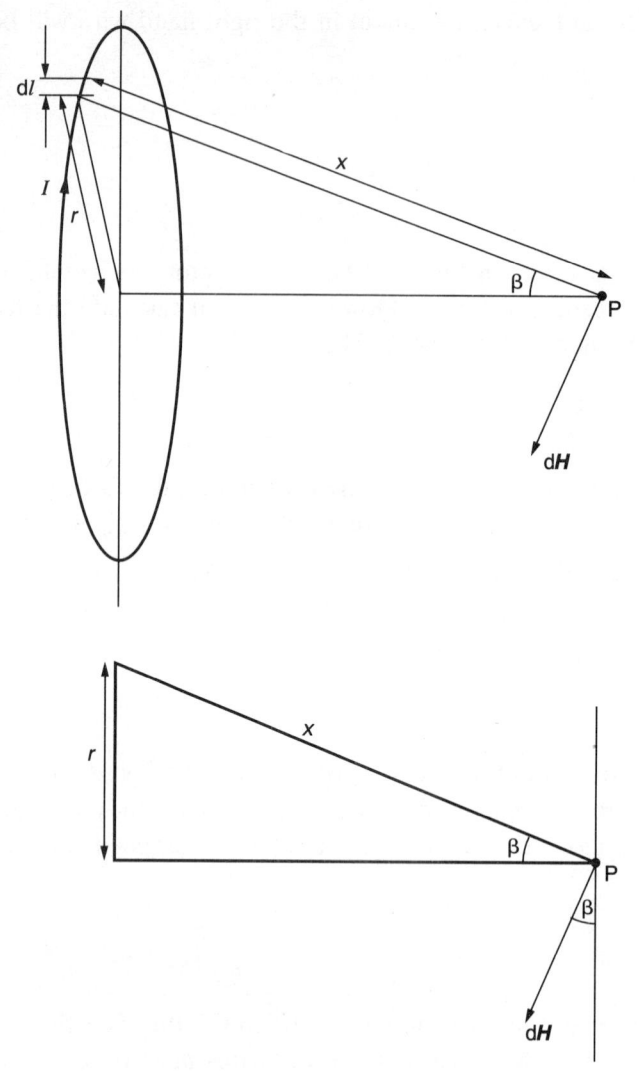

Fig. 3.9 Construction to find the magnetic field of a single-turn coil

Figure 3.9 shows a simple single-turn coil. We require to study the distribution of the magnetic field along (for simplicity) the axis of the coil. To analyse this situation, we will use the Biot–Savart law to find the field produced by a small section of the loop, and then integrate around the loop to find the total field.

Let us consider a simple current element of length dl. Now, from the Biot–Savart law, the magnitude of the magnetic field strength at point P is given by

$$dH = \frac{I\,dl}{4\pi x^2} \sin \theta$$

As the angle θ is $\pi/2$ in this instance, we can write

$$dH = \frac{I\,dl}{4\pi x^2} \tag{3.27}$$

Now, d**H** makes an angle to the horizontal of $\pi/2 - \beta$, and so we can resolve d**H** into vertical and horizontal components. When we integrate around the current loop, we find that the vertical component is zero due to the symmetry of the situation. (Interested readers can try this for themselves.) Thus we need only consider the horizontal component of d**H**, i.e. dH_p. So,

$$\begin{aligned} dH_p &= dH \cos(\pi/2 - \beta) \\ &= dH \sin\beta \\ &= \frac{I\,dl}{4\pi x^2}\sin\beta \end{aligned}$$

To find the total field, we need to integrate with respect to d*l*. When we do the integration, the only thing that varies is the elemental length d*l*. The limits of d*l* will be zero and $2\pi r$ (the circumference of the loop) and so the total field produced at point P is

$$\begin{aligned} H_p &= \frac{I\sin\beta}{4\pi x^2}\int_0^{2\pi r} dl \\ &= \frac{I\sin\beta}{4\pi x^2}\, 2\pi r \\ &= \frac{Ir}{2x^2}\sin\beta \end{aligned} \tag{3.28}$$

As $\sin\beta = r/x$, Equation (3.28) becomes

$$H_p = \frac{I\sin^3\beta}{2r}\ \text{A m}^{-1} \tag{3.29}$$

Although this equation gives the field along the axis of a single-turn coil, it is somewhat debatable whether it is of much use. After all, we seldom come across single-turn coils in real life, and even if we do, we hardly ever need to know what the field is. In spite of these remarks, this result is very important if we want to find the field due to a long coil of wire, or solenoid, and we often come across solenoids. This result is also used in physics when considering the field produced by Helmholtz coils, as the following example shows.

Example 3.3

A pair of coils, each consisting of *N* turns of wire and having a radius *r* metre, are situated in air. The coils face each other on the same axis, and they are separated by a distance equal to the radius of the coils. If each coil carries a

82 Electromagnetic fields

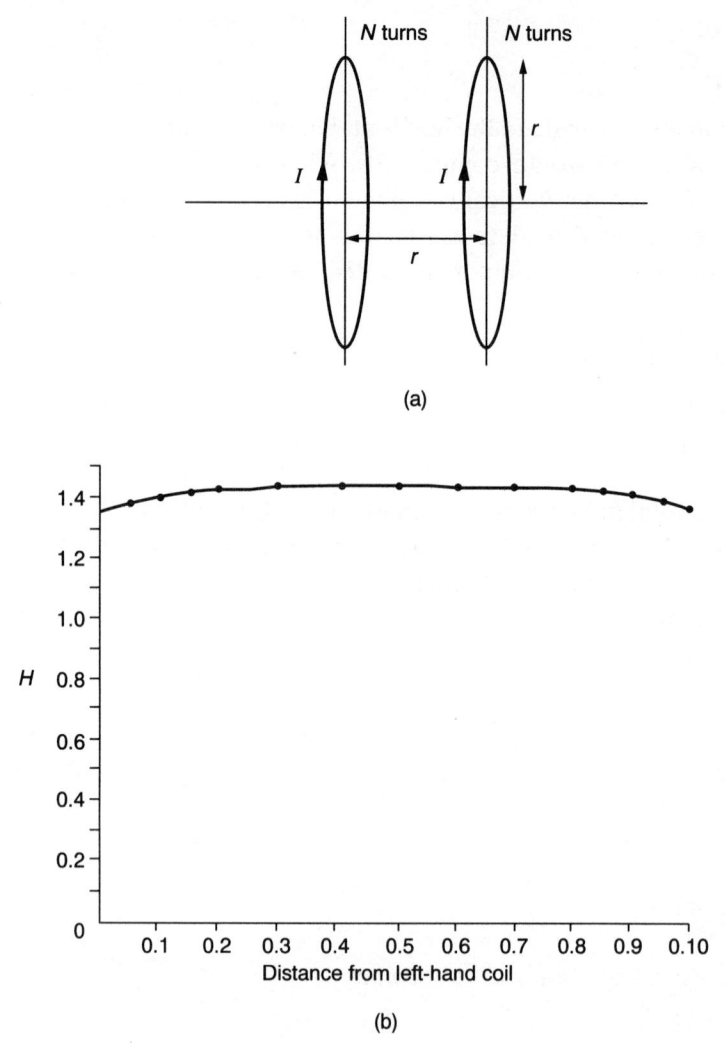

Fig. 3.10 (a) A pair of Helmholtz coils; and (b) variation in H between the two coils

current of I A in the same direction, plot the variation in magnetic field strength along the axis of the coils.

Solution

Figure 3.10(a) shows the situation we are to analyse. Now, to simplify the analysis, we will assume that the coils are very thin when compared with the distance between them. This means that we can assume the coils to be single turns carrying an effective current of NI amperes.

As we have just seen, the field along the axis of a single-turn coil is given by (Equation (3.29))

$$H = \frac{I \sin^3 \beta}{2r}$$

Thus the field due to the left-hand coil is

$$H_1 = \frac{NI \sin^3 \beta}{2r}$$

Now, $\sin \beta = \dfrac{r}{\sqrt{r^2 + x^2}}$ and so,

$$H_1 = \frac{NI}{2r} \frac{r^3}{(r^2 + x^2)^{3/2}}$$

$$= \frac{NI}{2} \frac{r^2}{(r^2 + x^2)^{3/2}} \quad \text{from the left-hand coil.}$$

As the right-hand coil is identical to the left-hand one, it will produce a field strength of

$$H_r = \frac{NI}{2} \frac{r^2}{(r^2 + (r - x)^2)^{3/2}}$$

giving a total field strength of

$$H = H_1 + H_r$$

$$= \frac{NI}{2} \frac{r^2}{(r^2 + x^2)^{3/2}} + \frac{NI}{2} \frac{r^2}{(r^2 + (r - x)^2)^{3/2}}$$

At this point we could apply the binomial theorem to simplify this equation. However, the result is rather complicated. Instead it is easier to plot the distribution of H directly, and Fig. 3.10(b) shows the result. As this plot shows, the variation of H, and hence B, is so small as to be neglected. Such coils are known as a Helmholtz pair (after Hermann von Helmholtz, 1821–1894, a German physicist).

3.9 THE SOLENOID

Figure 3.11 shows a long coil of wire, or solenoid. Such devices are often used as actuators with a bar magnet placed along the axis of the coil. Any current passing through the coil generates a magnetic field which forces the magnet in a particular direction. The magnet can then force a pair of contacts to close, or push a lever to move something.

To determine the field at any point along the axis of the solenoid, we will consider an elemental section of the coil, of thickness dx, and calculate the field produced. We will then integrate along the length of the coil to find the total field produced.

Let us assume that the solenoid has N turns and a length of l metre. With

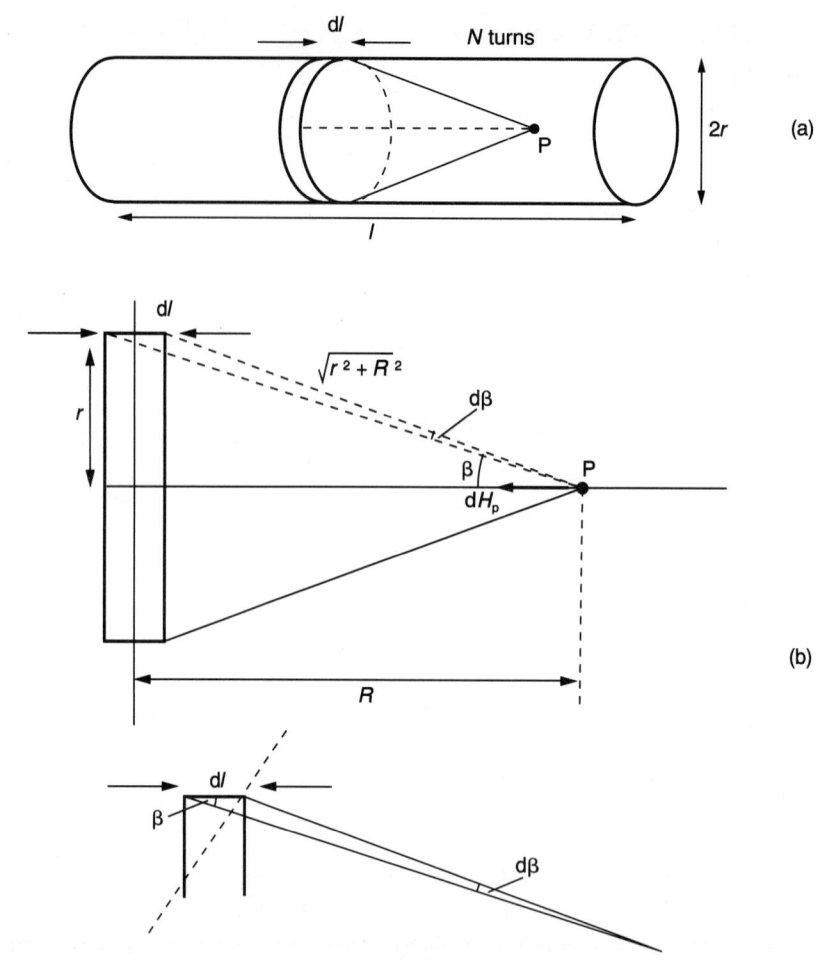

Fig. 3.11 (a) A simple solenoid; and (b) construction to determine the field produced by a solenoid

these figures, the number of turns per unit length will be N/l. Now, if we take a small section of the coil, of length dl, the number of turns in this section will be $dl \times N/l$. By using the result from the last section, the magnetic field strength generated by this section of the solenoid is

$$dH_p = dl \, \frac{N}{l} \, \frac{I \sin^3 \beta}{2r} \tag{3.30}$$

acting along the axis of the coil.

We now need to integrate along the length of the solenoid. However, as we move along the axis, the angle β changes between the limits β_{max} and β_{min}. So, we have to substitute for dl in terms of β. As Fig. 3.11(b) shows,

$$dl \sin \beta = d\beta \sqrt{r^2 + R^2}$$

and so,

$$dl = \frac{d\beta}{\sin \beta} \sqrt{r^2 + R^2}$$

Thus Equation (3.30) becomes

$$dH_p = \frac{d\beta}{\sin \beta} \sqrt{r^2 + R^2} \frac{N}{l} \frac{I \sin^3 \beta}{2r}$$

$$= \frac{\sqrt{r^2 + R^2}}{2r} \frac{N}{l} I \sin^2 \beta \, d\beta \qquad (3.31)$$

Now, $\sin \beta = \dfrac{r}{\sqrt{r^2 + R^2}}$ and so we can write

$$dH_p = \frac{NI}{2l} \sin \beta \, d\beta$$

Thus the total field at point P is

$$H_p = \frac{NI}{2l} \int_{\beta_{max}}^{\beta_{min}} \sin \beta \, d\beta$$

$$= \frac{NI}{2l} (\cos \beta_{min} - \cos \beta_{max}) \text{ A m}^{-1} \qquad (3.32)$$

We can use this equation to find the field at the centre of the solenoid. If P is at the centre, $\beta_{max} = 180° - \beta_{min}$, and so

$$H_p = \frac{NI}{2l} (\cos \beta_{min} - \cos (180° - \beta_{max}))$$

$$= \frac{NI}{2l} (\cos \beta_{min} + \cos \beta_{min})$$

$$= \frac{NI}{l} \cos \beta_{min} \text{ A m}^{-1} \qquad (3.33)$$

If the solenoid is very long, $\beta_{min} \approx 0$, and so the field at the centre of the coil approximates to

$$H_p = \frac{NI}{l} \text{ A m}^{-1} \qquad (3.34)$$

Example 3.4

A 2 cm diameter solenoid is 10 cm long, and has 30 turns per cm. When energized, the solenoid takes 2 A of current. Plot the variation in H and B along the axis of the coil. (Assume that the solenoid is air-cored.)

86 Electromagnetic fields

If a bar magnet of strength 10 mWb, and length 5 cm, is placed in the solenoid, plot the variation in force along the axis when the coil is energized.

Solution

We require to find the variation in H and B along the axis of the solenoid. Now, Equation (3.32) gives the variation in H as

$$H_p = \frac{NI}{2l}(\cos \beta_{min} - \cos \beta_{max}) \text{ A m}^{-1}$$

and so,

$$H_p = \frac{30 \times 10 \times 2}{2 \times 0.1}(\cos \beta_{min} - \cos \beta_{max})$$

$$= 3 \times 10^3 (\cos \beta_{min} - \cos \beta_{max}) \text{ A m}^{-1}$$

As the solenoid is air-cored, $\mu = 4\pi \times 10^{-7}$ and so

$$B_p = 3.8 \times 10^{-3} (\cos \beta_{min} - \cos \beta_{max}) \text{ Wb m}^{-2}$$

Figure 3.12(a) shows the variation in H and B along the axis of the coil. As can be seen, the maximum field occurs at the centre of the solenoid.

Now, if we place a 10 mWb bar magnet in the solenoid, the magnet will experience a force of

$$F = 10 \times 10^{-3} H_p$$

at each pole. As the magnet has a length of 5 cm, the two poles will experience different forces due to the variation in H as we move along the solenoid axis. Thus it is a question of calculating the force on each pole, and adding the two results to get the total force. As a sample calculation, we will take the left-hand pole of the bar magnet to be at the centre of the solenoid.

So, the H field at the centre is

$$H = \frac{NI}{l} \cos \beta_{min}$$

$$= \frac{30 \times 10 \times 2}{0.1} \cos \beta_{min}$$

$$= 6 \times 10^3 \times \frac{5}{\sqrt{5^2 + 1^2}}$$

$$= 5.9 \times 10^3 \text{ A m}^{-1}$$

Thus the force on the left-hand pole is

$$F = 5.9 \times 10^3 \times 10 \times 10^{-3}$$
$$= 59 \text{ newton}$$

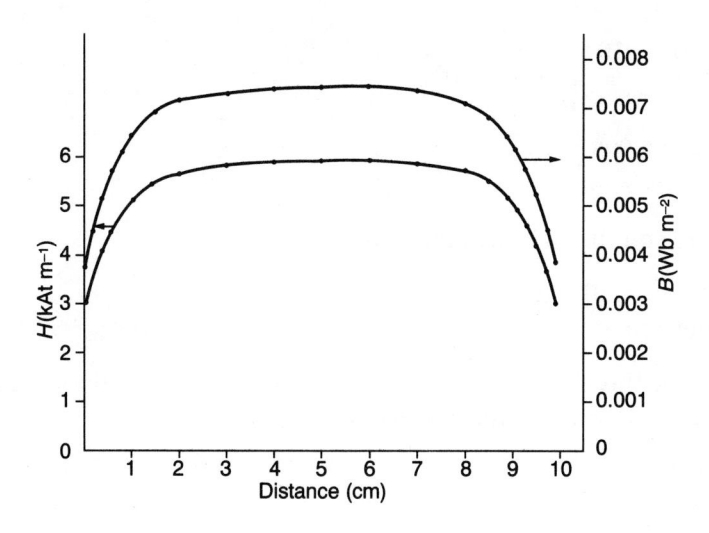

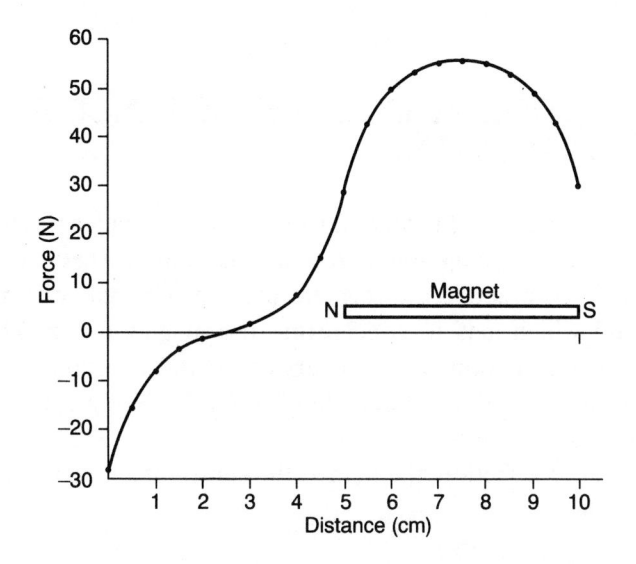

Fig. 3.12 (a) Variation of B and H along the axis of a solenoid; and (b) variation of force along the axis of a solenoid

The length of the magnet is 5 cm, and so the H field 5 cm to the right of the solenoid centre is

$$H = \frac{NI}{2l}(\cos\beta_{min} - \cos\beta_{max})$$

$$= \frac{30 \times 10 \times 2}{2 \times 0.1}(\cos\beta_{min} - \cos\beta_{max})$$

$$= 3 \times 10^3 \left\{ \frac{10}{\sqrt{10^2 + 1^2}} - \cos 90° \right\}$$

$$= 3 \times 10^3 \times 0.995$$
$$= 2.985 \times 10^3 \text{ A m}^{-1}$$

Thus the force on the right-hand pole is

$$F = 2.985 \times 10^3 \times 10 \times 10^{-3}$$
$$= 29.85 \text{ newton}$$

As the pole at this location is the opposite of the pole at the centre of the solenoid, the total force on the bar magnet is

$$F_t = 59 - 29.85$$
$$= 29.15 \text{ newton}$$

Figure 3.12(b) shows the variation in this force as the magnet moves along the axis of the solenoid.

3.10 THE TOROIDAL COIL, RELUCTANCE AND MAGNETIC POTENTIAL

Figure 3.13 shows the general form of a toroidal former which has a coil wound on it. This is basically a long solenoid which is bent so that the coil has no beginning or end. In practice the formers used in toroidal coils are made of powdered ferrite which acts to concentrate the magnetic flux. Thus leakage effects are minimal, so making the coil very efficient. This is put to good use in transformers, which we will encounter in Chapter 7. Here we want to develop our model of electromagnetism further.

As we saw in the last section, the field at the centre of a long solenoid is given by (Equation (3.34))

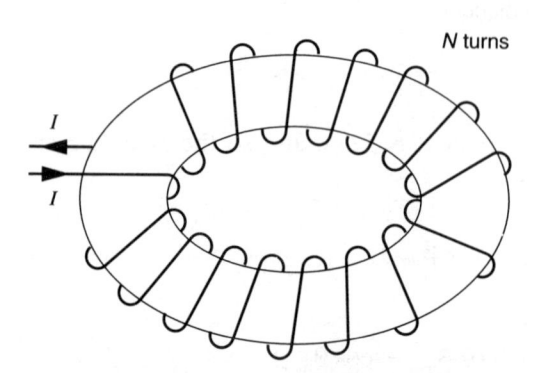

$$H = \frac{NI}{l} \text{ A m}^{-1} \tag{3.35}$$

Fig. 3.13 A typical toroidal coil

3.10 The toroidal coil, reluctance and magnetic potential

where l is the length of the solenoid. As the coil is wound on a toroid, this field will be constant along the length of the coil. As the coil is circular, the length of the solenoid will be the average circumference of the former. Now, as $B = \mu H$, the flux density in the former will be

$$B = \mu \frac{NI}{l} \text{ Wb m}^{-1} \tag{3.36}$$

with μ being the permeability of the former. As B is the flux density, Equation (3.36) becomes

$$\frac{\phi}{\text{area}} = \mu \frac{NI}{l}$$

and so,

$$NI = \phi \times \frac{1}{\mu \times \text{area}} \tag{3.37}$$

Let us examine this equation closely. The first term on the right-hand side of this equation is the magnetic flux that flows around the toroid. The second term is similar to our formula for the capacitance of a parallel plate capacitor, Equation (2.35). As we saw in Section 2.9, we can regard capacitance as a measure of the resistance to the flow of flux. So, could we take this second term as a measure of resistance to magnetic flux?

Let us define the reluctance as

$$S = \frac{l}{\mu \times \text{area}} \tag{3.38}$$

with units of At Wb^{-1}, i.e. ampere-turns per Weber. (Most texts choose not to use units of turns as they are not part of the SI system. However, the use of ampere-turns is helpful when dealing with electromagnetism, and so we will retain them.) As Equation (3.38) shows, the reluctance is directly dependent on the length of the coil. Thus if the toroidal former has a large circumference, the reluctance is high, and it becomes more difficult to produce flux in the core. If we choose to use a former with a high permeability, the reluctance will be low, and so it will become easier to produce flux in the core. Thus we can regard the reluctance as the resistance to the flow of flux.

To return to Equation (3.37), the term on the left-hand side is the number of turns times the current flowing through them. If we continue drawing our parallel with electrostatics, the ampere-turns appear to be the equivalent of potential. We defined potential as the work done in moving a unit charge through an electric field. So, are the ampere-turns equal to the work done in moving a unit pole through a magnetic field?

As we saw at the start of this chapter, the force on a unit pole placed in a magnetic field is (Equation (3.3))

$$F = p_2 H$$

Now, the magnetic field in the core is (Equation (3.35))

$$H = \frac{NI}{l} \text{ A m}^{-1}$$

or,

$$H = \frac{NI}{2\pi r} \text{ A m}^{-1} \qquad (3.39)$$

where r is the average radius of the former. So, if we move a unit north pole around the former, we have to do work of

work done = force × distance

$$= 1 \times \frac{NI}{2\pi r} \times 2\pi r$$

$$= NI \text{ At}$$

Thus the magnetic potential (the work done) is given by

$$V_m = NI \text{ ampere-turns} \qquad (3.40)$$

We can now rewrite Equation (3.37) as

$$V_m = \phi \times S \qquad (3.41)$$

So, the magnetic potential is equal to the product of the magnetic flux and reluctance. This is very similar to Ohm's law which applies to electroconductive fields. We will put this similarity to good use when we consider transformers in Chapter 7. (Some texts refer to the magnetic potential as the magneto-motive force, or mmf. However, this gives an image of the mmf forcing the flux around the magnetic circuit. As we have seen, the mmf is not a force, but rather a measure of the energy required to move a pole around a magnetic field. As the term mmf can be a source of confusion, we will refer to it as magnetic potential.)

Example 3.5

A toroidal coil is wound on a ferrite former with inner radius 4 cm, and outer radius 6 cm. The coil has 2000 turns, and the core has a relative permeability of 200. Determine the current required to produce a mean flux of 1 mWb in the core.

Solution

The magnetic potential is the product of the flux and the reluctance of the core. Now, before we can find the reluctance, we need to find the average circumference and cross-sectional area of the core. The average radius of the core is

$$\text{average radius} = \frac{4+6}{2} \text{ cm}$$
$$= 5 \text{ cm}$$

Thus the average length of the magnetic path is

$l = 2\pi \times 5 \times 10^{-2}$
$= 0.314 \text{ m}$

The former has a circular cross-sectional area of radius

$$r = \frac{6-4}{2} \text{ cm}$$
$= 1 \text{ cm}$

and so the reluctance of the core is (Equation (3.38))

$$S = \frac{1}{\mu \times \text{area}}$$

$$= \frac{0.314}{200 \times 4\pi \times 10^{-7} \times \pi \times (1 \times 10^{-2})^2}$$
$= 4 \times 10^6 \text{ At Wb}^{-1}$

Thus the ampere-turns required are (Equation (3.41))

$V_m = 1 \times 10^{-3} \times 4 \times 10^6$
$= 4 \times 10^3 \text{ At}$

As the coil has 2000 turns, the coil current is

$$I = \frac{4 \times 10^3}{2000}$$

$= 2 \text{ amperes}$

We should note that, in reality, the distribution of magnetic flux across the cross-sectional area of the toroid is uneven. In spite of this, this derivation does give an indication of the required current.

3.11 INDUCTANCE

So far we have only considered coils that have a steady d.c. current passing through them. This introduced us to the idea of magnetic flux, magnetic flux density, and magnetic field strength. Although d.c. circuits sometimes use coils, we more usually find them in a.c. circuits. In such circuits, we tend to characterize coils by a term called inductance.

When a d.c. voltage energizes a coil, a current flows which sets up a magnetic field around the coil. This field will not appear instantaneously as it takes a certain amount of time to produce the field. After the initial transient has passed, the resistance of the wire that makes up the coil will limit its current.

Let us now consider a very low-resistance coil connected to a source of alternating voltage. As the coil resistance is very low, the coil should appear to be a short-circuit. This should result in a lot of current flowing! However, what we find is that the current taken by the coil depends on the frequency of the source – high frequencies result in low currents. Thus some unknown property of the coil restricts the current.

In 1831, a British physicist, chemist and great experimenter called Michael Faraday (1791–1867) was investigating electromagnetism. As a result of his experiments, Faraday proposed that a changing magnetic field induces an emf into a coil. This was one of the most significant discoveries in electrical engineering, and it is the basic principle behind transformers and electrical machines. (Faraday's achievement is even more remarkable in that all of his work resulted from experimentation, and not mathematical derivation.)

Faraday's law formalizes this result as

$$e \propto \frac{d}{dt}(N\phi) \tag{3.42}$$

where N is the number of turns in the coil, and $N\phi$ is known as the flux linkage. So, the induced emf depends on the rate of change of flux linkages, i.e. the higher the frequency, the higher the rate of change, the larger the induced emf. As this emf serves to oppose the voltage that produces it, Equation (3.42) is often modified to

$$e = -\frac{d}{dt}(N\phi) \tag{3.43}$$

So, if we have a coil connected to a source of alternating voltage, the coil takes current which produces a magnetic field. This magnetic field produces an alternating magnetic flux, by virtue of $B = \mu H$, so generating a back-emf in the coil (note the minus sign in Equation (3.43)). This back-emf opposes the voltage producing it, and this reduces the current taken from the supply.

An alternative 'circuits' expression for the back-emf is

$$e = -\frac{d}{dt}(Li) \tag{3.44}$$

where L is the inductance of the coil, and i is the alternating current taken by the coil. Now, by equating these two expressions we have

$$-\frac{d}{dt}(N\phi) = -\frac{d}{dt}(Li)$$

and so,

$$L = N\frac{d\phi}{di} \tag{3.45}$$

Thus the inductance is the flux linkage per unit current. The unit of inductance is the Henry, named after the American physicist Joseph Henry (1791–1878) who invented the electromagnetic telegraph as well as extensively studying electromagnetic phenomena.

We will now go on to examine the inductance of a simple coil, which may be air-cored or iron-cored. However, what is seldom appreciated is that a piece of wire can have an inductance – termed self-inductance. Then we will also examine the inductance of some transmission lines.

Simple coil

A simple coil consists of several turns of wire wound around a former. As we have just seen, the inductance is defined as the flux linkage per unit current, i.e.

$$L = N\frac{d\phi}{di}$$

where N is the number of turns in the coil. When we considered solenoids, we saw that the flux density varies along the axis of the coil. However, if the coil is very long, the field at the centre of the coil is

$$H = \frac{NI}{l}$$

and so,

$$B = \mu\frac{NI}{l}$$

As B is the flux density, i.e. $B = \phi/A$, we can write

$$L = N\frac{dB}{di}A$$

$$= NA\mu\frac{N}{l}\frac{di}{di}$$

$$= \frac{N^2\mu A}{l} \tag{3.46}$$

where A is the cross-sectional area of the coil. Although Equation (3.46) gives the inductance of a long coil, this equation is an approximation. This is because it assumes that the field is constant throughout the coil, and it neglects the effects of flux leakage.

Example 3.6

An air-cored coil has a diameter of 2 cm, and a length of 10 cm. The number of turns per cm is 50. Estimate the inductance of the coil. If a powdered ferrite core is added, with μ_r of 100, estimate the new value of inductance. Determine the new coil length if the ferrite-core coil is to have the same value of inductance as the original air-cored coil.

Solution

The coil is 10 cm long, and has 50 turns per cm. So, the total number of turns on the coil is 500. From Equation (3.46), the inductance is

$$L = \frac{500^2 \times 4\pi \times 10^{-7} \times \pi \times (1 \times 10^{-2})^2}{10 \times 10^{-2}}$$

$$= 1 \text{ mH}$$

If we now insert a ferrite core the value of μ will increase. As the core has $\mu_r = 100$, the new inductance is

$$L = 1 \times 10^{-3} \times 100$$
$$= 100 \text{ mH}$$

We now require the iron-cored coil to have the same inductance as the original air-cored coil. To do this we will retain the same number of turns per metre (5×10^3) and reduce the length of the coil. So,

$$L = \frac{(5 \times 10^3 \times 1)^2 \times \mu_0\mu_r \times A}{1} = 1 \text{ mH}$$

After some rearranging, this gives

$$l = \frac{1 \times 10^{-3}}{(5 \times 10^3)^2 \times \mu_0\mu_r \times A}$$

$$= 1 \times 10^{-3} \text{ m}$$
$$= 1 \text{ mm}$$

Thus the number of turns is only 5, and the length of the coil is 1 mm. This means that the coil length is much smaller than the diameter, and so we must treat the calculated inductance value with suspicion. However, this example does show the effects of a ferrite core on the length of a coil.

Self-inductance of a single wire

We have just seen that inductance is equal to the flux linkages per unit current. The example at the end of Section 3.6 examined the variation of magnetic field both inside and outside the wire. As we saw, the magnetic field inside the wire is associated with a fraction of the current flowing through the wire. So, as we

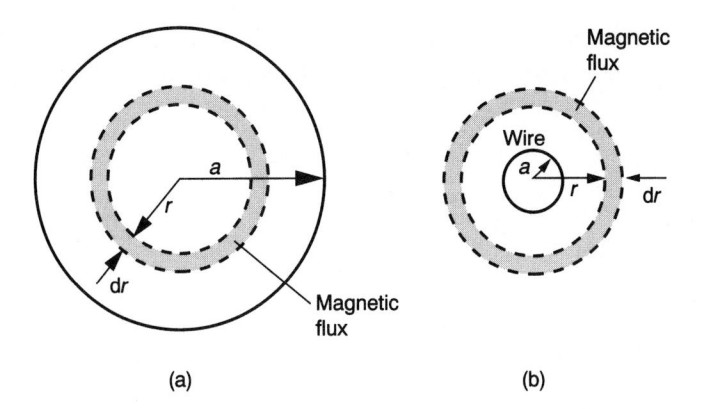

Fig. 3.14 (a) Cross-section of a current-carrying wire; and (b) magnetic flux outside a current-carrying wire

have different fields inside and outside the wire, we should expect that there are two components to the inductance of a piece of wire.

Let us initially examine the field inside the wire. Figure 3.14(a) shows a cross-section of the wire, in which we have a circular path of radius r and thickness dr. If we assume a constant current density, the current enclosed by this loop is

$$I' = \frac{I}{\pi a^2} \pi r^2$$

This current will produce a magnetic field of

$$H_r = \frac{I}{\pi a^2} \pi r^2 \cdot \frac{1}{2\pi r}$$

$$= \frac{Ir}{2\pi a^2}$$

which gives a flux density of

$$B_r = \frac{\mu_0 I r}{2\pi a^2} \quad \text{at a radius of } r$$

We require an expression linking the flux, at radius r, to the current producing the flux. So, in order to find the flux, we will consider an incremental ring of thickness dr and radius r as shown in Fig. 3.14(a). The area that the flux sees is $dr \times$ length, and so the flux through this ring is

$$d\phi = B_r \times dr \times \text{length}$$

$$= \frac{\mu_0 I r}{2\pi a^2} dr \times \text{length}$$

This flux is generated by the fraction of the total current flowing through the wire. Now, in order to find the fractional flux linkage, we have to know how much of the wire is linked with this flux. Thus we need to multiply the incremental flux by a fraction. Specifically, the fractional flux linkage, $d\ddot{U}$, is

$$d\ddot{U} = d\phi \times \frac{\pi r^2}{\pi a^2}$$

$$= \frac{\mu_0 I r}{2\pi a^2} dr \times \text{length} \times \frac{\pi r^2}{\pi a^2}$$

$$= \frac{\mu_0 I}{2\pi a^4} r^3 \times dr \times \text{length}$$

So, the total flux linkage inside the wire is

$$\ddot{U} = \frac{\mu_0 I}{2\pi a^4} \int_0^a dr \times \text{length}$$

$$= \frac{\mu_0 I}{2\pi a^2} \left(\frac{r^4}{4}\right)_0^a \times \text{length}$$

$$= \frac{\mu_0 I}{8\pi a^4} \frac{a^4}{4} \times \text{length}$$

$$= \frac{\mu_0 I}{8\pi} \times \text{length}$$

Hence the total internal inductance per unit length is

$$L' = \frac{\mu_0}{8\pi} \text{ H m}^{-1} \tag{3.47}$$

This is the internal inductance of the wire. As this equation shows, it is independent of the diameter of the wire.

Let us now turn our attention to the flux outside the wire. Application of Ampère's law gives the field strength at a radius r as

$$H_r = \frac{I}{2\pi r}$$

and so the flux density at this radius is

$$B_r = \frac{\mu_0 I}{2\pi r}$$

Let us again consider an incremental tube of radius r and thickness dr – see Fig. 3.14(b). The fractional flux through the surface of this ring is

$$d\phi = B_r \times dr \times \text{length}$$

$$= \frac{\mu_0 I}{2\pi r} dr \times \text{length}$$

This is the fractional flux linkage generated by the total current flowing in the wire. So, the fractional inductance per unit length is

$$dL' = \frac{d\phi}{I}$$

$$= \frac{\mu_0}{2\pi} \frac{dr}{r}$$

and the total external inductance per unit length is

$$L' = \frac{\mu_0}{2\pi} \int_a^\infty \frac{dr}{r}$$

$$= \frac{\mu_0}{2\pi} \left| \ln r \right|_a^\infty = \frac{\mu_0}{2\pi} (\ln \infty - \ln a)$$

$$= \text{infinity} \tag{3.48}$$

This rather surprising result arises because we assumed that the field is zero at infinity. (A similar situation arose when we examined line charges in Section 2.6.) However, we should remember that there must be a return path for the current – it may be a very long way away, but it must be there. What usually happens is that the wire has a ground-plane at a certain distance from the conductor. We will meet this situation again when we consider the inductance of microstrip later in this section.

Example 3.7

Determine the inductance per unit length of 5 mm diameter, copper wire.

Solution

We have wire of diameter 5 mm, and so the radius is 2.5 mm. The internal self-inductance of the wire is independent of radius, and so

$$L_{int} = \frac{\mu_0}{8\pi}$$

$$= \frac{4\pi \times 10^{-7}}{8\pi}$$

$$= 50 \text{ nH}$$

As regards the external inductance, we can write

$$L_{ext} = \frac{\mu_0}{2\pi} (\ln d - \ln a)$$

Let us calculate the total inductance assuming the magnetic field is zero at certain distances.

Distance	External inductance (nH)	Total inductance (nH)	Percentage increase
$2a$	140	190	–
$20a$	600	650	242
$200a$	1060	1110	71
$2000a$	1520	1570	41
$2 \times 10^4 a$	2000	2050	30
$2 \times 10^5 a$	2440	2490	21

This table shows that the percentage increase in total inductance falls as the distance from the wire increases. We should also note that the internal inductance becomes insignificant when compared with the external inductance.

Coaxial cable

Figure 3.15 shows a cross-section through a length of coaxial cable. Now, current in the inner conductor generates a magnetic field in the inner conductor, and in the dielectric between the inner and outer conductors. The outer conductor is usually earthed, and effectively shields the signal on the inner conductor from any external interference. Thus the field at the outer conductor is zero.

As we have seen in the previous section, there will be two parts to the inductance: the inductance of the inner conductor; and the inductance due to the magnetic field in the dielectric. As regards the inductance of the inner conductor, the last section gave a value of

$$L_{int} = \frac{\mu_0}{8\pi} \text{ H m}^{-1}$$

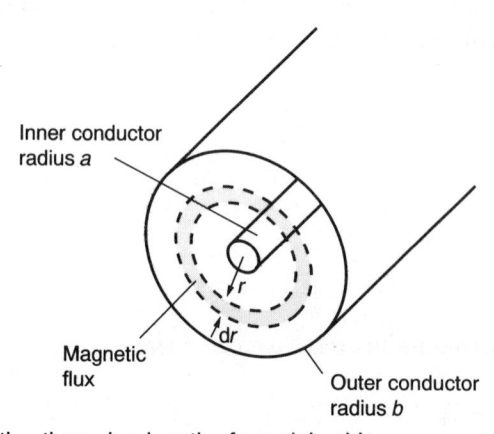

Fig. 3.15 Cross-section through a length of coaxial cable

To find the external inductance, we will follow a similar procedure to that used in the last section. Thus the flux density at a radius r in the dielectric is

$$B_r = \mu_0 \frac{I}{2\pi r} \tag{3.49}$$

With this flux density, the flux through a small incremental ring of radius r and thickness dr is

$$d\phi = B_r \times dr \times \text{length}$$

$$= \frac{\mu_0 I}{2\pi r} dr \times \text{length}$$

This is the fractional flux linkage generated by the current I flowing in the inner conductor. Thus the fractional external inductance per unit length is

$$dL' = \frac{d\phi}{I}$$

$$= \frac{\mu_0}{2\pi} \frac{dr}{r}$$

and so the total external inductance per unit length is

$$L_{\text{ext}} = \frac{\mu_0}{2\pi} \int_a^b \frac{dr}{r}$$

$$= \frac{\mu_0}{2\pi} \left| \ln r \right|_a^b$$

$$= \frac{\mu_0}{2\pi} (\ln b - \ln a)$$

$$= \frac{\mu_0}{2\pi} \ln(b/a) \tag{3.50}$$

Thus the total inductance per unit length is

$$L' = \frac{\mu_0}{8\pi} + \frac{\mu_0}{2\pi} \ln(b/a) \tag{3.51}$$

We should note that the internal inductance can be insignificant when compared with the external inductance. Also, at high frequencies the current tends to crowd towards the surface of the inner conductor. This is known as the skin effect, and when it occurs, the internal inductance tends to zero. Thus the inductance per unit length approximates to

$$L' = \frac{\mu_0}{2\pi} \ln(b/a) \text{ H m}^{-1} \tag{3.52}$$

As we saw in Section 2.9, such a cable also has a capacitance given by (Equation (2.37))

$$C' = \frac{2\pi\epsilon_0\epsilon_r}{\ln(b/a)} \text{F m}^{-1}$$

Now, if we multiply these two equations together, we get

$$L'C' = \frac{\mu_0}{2\pi}\ln(b/a)\frac{2\pi\epsilon_0\epsilon_r}{\ln(b/a)}$$

$$= \mu_0\,\epsilon_0\epsilon_r \tag{3.53}$$

So, the product of inductance and capacitance results in the product of permeability and permittivity. This is an interesting result, which we will return to in the next section.

Example 3.8

A 500 m length of coaxial cable has an inner conductor of radius 2 mm, and an outer conductor of radius 1 cm. A non-ferrous dielectric separates the two conductors. Determine the inductance of the cable.

Solution

The dielectric is non-ferrous, and so the permeability of the dielectric is that of free space. So, the inductance per unit length is (Equation (3.51))

$$L' = \frac{\mu_0}{8\pi} + \frac{\mu_0}{2\pi}\ln(b/a)$$

$$= 50 \times 10^{-9} + 200 \times 10^{-9}\ln 5$$
$$= 50 \times 10^{-9} + 322 \times 10^{-9}$$
$$= 372 \text{ nH per metre}$$

As the cable is 500 m long, the inductance is

$L = 372 \times 500$ nH
$ = 186\ \mu\text{H}$

At frequencies above about 1 MHz, we can neglect the internal inductance. Thus the inductance of the cable becomes

$L = 161\ \mu\text{H}$

The dimensions of this cable are the same as the cable in Example 2.9 used in Section 2.9 in the last chapter.

Twin feeder

Figure 3.16 shows a section through a length of twin feeder. The left-hand conductor carries a current of I A, while the return conductor (the right-hand one) carries a current of $-I$ A. Both conductors will produce magnetic fields in

3.11 Inductance

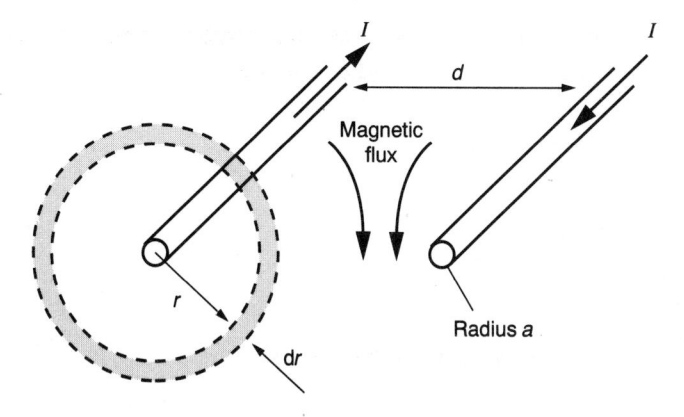

Fig. 3.16 Cross-section through a length of twin feeder

the space between the wires, and so there will be two components to the flux density at any particular point.

As before, there will be two components to the inductance: the internal inductance of each conductor; and the external inductance of each conductor. We have already calculated the internal inductance of a conductor. However, to find the external inductance, we need to generate an equation linking the total flux with the current in one of the conductors.

To analyse the situation, we will use the principle of superposition to find the total flux density at some point between the two wires. So, the flux density at a radius r due to the current in the left-hand conductor is

$$B_1 = \frac{\mu_0 I}{2\pi r} \tag{3.54}$$

The right-hand conductor generates a flux density at the same point of

$$B_r = \frac{\mu_0 I}{2\pi(d-r)} \tag{3.55}$$

acting in the same direction. So, the total field at a radius of r from the left-hand conductor is

$$B_t = \frac{\mu_0 I}{2\pi r} + \frac{\mu_0 I}{2\pi(d-r)}$$

Thus the fractional flux through an incremental ring of radius r and thickness dr is

$$d\phi = \left(\frac{\mu_0 I}{2\pi r} + \frac{\mu_0 I}{2\pi(d-r)}\right) dr \times \text{length}$$

and so the fractional inductance per unit length is

$$dL' = \left(\frac{\mu_0}{2\pi r} + \frac{\mu_0}{2\pi(d-r)}\right) dr \tag{3.56}$$

In order to find the total inductance per unit length, we need to integrate this equation between the limits a and $d - a$. So,

$$L' = \frac{\mu_0}{2\pi} \int_a^{d-a} \left(\frac{1}{r} + \frac{1}{d-r} \right) dr$$

$$= \frac{\mu_0}{2\pi} \left| \ln r - \ln(d-r) \right|_a^{d-a}$$

$$= \frac{\mu_0}{2\pi} (\ln(d-a) + \ln a + \ln a + \ln(d-a))$$

$$= \frac{\mu_0}{\pi} \ln \left(\frac{d-a}{a} \right) \text{ H m}^{-1} \tag{3.57}$$

As we have two conductors, the internal inductance, from our previous studies, is doubled. Thus,

$$L' = \frac{\mu_0}{4\pi} \text{ H m}^{-1}$$

and so the total inductance per unit length of twin feeder is

$$L' = \frac{\mu_0}{4\pi} + \frac{\mu_0}{\pi} \ln \left(\frac{d-a}{a} \right) \text{ H m}^{-1} \tag{3.58}$$

If we can ignore the internal inductance, we get

$$L' = \frac{\mu_0}{\pi} \ln \left(\frac{d-a}{a} \right) \text{ H m}^{-1} \tag{3.59}$$

As we saw in Section 2.9, this type of cable also has capacitance given by (Equation (2.44))

$$C' = \frac{\pi \epsilon_0}{\ln((d-a)/a)} \text{ F m}^{-1}$$

If we multiply these two equations together, we get

$$L'C' = \frac{\mu_0}{\pi} \ln \left(\frac{d-a}{a} \right) \frac{\pi \epsilon_0}{\ln((d-a)/a)}$$

$$= \mu_0 \epsilon_0$$

This is the same result as that obtained with coaxial cable, Equation (3.53). This has the making of a rule:

> In a transmission line, the product of capacitance and inductance equals the product of the permittivity and permeability, i.e.

$$L'C' = \mu\epsilon$$

As the speed of light is $1/\sqrt{\mu\epsilon}$, we can write

3.11 Inductance

$$c = \frac{1}{\sqrt{L'C'}} = \frac{1}{\sqrt{\mu\epsilon}} \tag{3.60}$$

This shows that signals propagate down the line at the speed of light in the dielectric (slower than they travel in free space). We will use this rule when we consider microstrip in the next section.

Example 3.9

A 200 m length of feeder consists of two 2 mm radius conductors separated by a distance of 20 cm. Determine the inductance of the arrangement.

Solution

If we assume that the conductors are in air, the permeability of the dielectric surrounding them is that of free space. Thus, from Equation (3.58) we get

$$\begin{aligned} L' &= 100 \times 10^{-9} + 400 \times 10^{-9} \ln 99 \\ &= 100 \times 10^{-9} + 400 \times 10^{-9} \times 4.6 \\ &= 100 \times 10^{-9} + 1840 \times 10^{-9} \\ &= 1.94 \; \mu\text{H m}^{-1} \end{aligned}$$

So, the total inductance of the cable is

$$\begin{aligned} L &= 1.94 \times 10^{-6} \times 200 \\ &= 388 \; \mu\text{H} \end{aligned}$$

At high frequencies, the internal inductance tends to zero, and so the cable inductance tends to

$$L = 368 \; \mu\text{H m}^{-1}$$

The dimensions of this cable are identical to that used in Example 2.10 in Section 2.9. So, twin feeder has inductance and capacitance.

Microstrip lines

Let us now consider microstrip. This was first introduced in Section 2.9 when we considered the capacitance of a printed circuit track over a ground-plane. To find the inductance of this arrangement, we would have to plot the magnetic field surrounding the track. This involves a considerable amount of work, which we can avoid by using the rule introduced at the end of the last section.

When we considered coaxial cable and twin feeder, we found that

$$L'C' = \mu\epsilon$$

So, if we can determine the capacitance, we can find the inductance per unit length. In Section 2.9 we found that we could approximate the capacitance to that of a parallel plate capacitor (Equation (2.48)):

$$c' = \frac{\epsilon_0 \epsilon_r w}{h} \text{ F m}^{-1}$$

or, if the track width is much less than the thickness of the board, the capacitance of a cylindrical wire over a ground-plane (Equation (2.49)):

$$C' = \frac{2\pi\epsilon_0\epsilon_r}{\ln(h/w)} \text{ F m}^{-1}$$

Thus the inductance lies between

$$L' = \frac{\mu_0 h}{w} \text{ H m}^{-1} \tag{3.61}$$

for the parallel plate approximation, and

$$L' = \frac{\mu_0}{2\pi} \ln(h/w) \text{ H m}^{-1} \tag{3.62}$$

for the wire above ground approximation.

Example 3.10

A 3 mm wide track is etched on one side of some double-sided printed circuit board. The thickness of the board is 2 mm, and the dielectric has a relative permeability of 1. Determine the inductance per cm.

Solution

As the track width is of the same order of magnitude as the board thickness, we must use Equation (3.61) to give

$$L' = \frac{\mu_0 h}{w} \text{ H m}^{-1}$$

$$= 4\pi \times 10^{-7} \frac{2 \times 10^{-3}}{3 \times 10^{-3}}$$

$$= 8.4 \times 10^{-7} \text{ H m}^{-1}$$

$$= 8.4 \text{ nH cm}^{-1}$$

This neglects the effect of internal inductance given by

$$L' = \frac{\mu_0}{8\pi}$$

$$= 50 \text{ nH m}^{-1}$$

$$= 0.5 \text{ nH cm}^{-1}$$

Thus the total inductance is

$$L' = 8.9 \text{ nH cm}^{-1}$$

Energy storage

We are generally familiar with the storage of energy in capacitors – computer memory boards often use them as back-up power supplies. However, we seldom come across coils as energy storage devices.

One application where coils are used as energy storage devices is in car ignition systems. In a conventional ignition system, the low-tension, LT, part of the coil is energized with 12 volt. When the contact breakers open, they disconnect the LT coil from the car battery, and so the magnetic field collapses. This collapsing field generates a changing magnetic flux, which induces a voltage into the high-tension, HT, coil which produces the spark across the contacts in the spark plug. According to the principle of conservation of energy, the energy stored in the magnetic field must appear as energy across the spark plug contacts. So, the magnetic field stores energy.

To find the stored energy, let us take an inductor connected to a d.c. source. This inductor will take a certain amount of current, limited by the resistance of the coil. If we increase the current by a small amount dI in time dt, the flux causes a back-emf given by

$$dV = L\frac{dI}{dt} \tag{3.63}$$

As the current flowing through the coil is I, the instantaneous power supplied is

$$I\,dV = LI\frac{dI}{dt} \tag{3.64}$$

This power is supplied in time dt, and so the energy supplied in raising the current from I to $I + dI$ is

$$dE = I\,dV\,dt$$

$$= LI\frac{dI}{dt}dt$$

$$= LI\,dI \tag{3.65}$$

Thus we can find the energy supplied in raising the current from zero to I by integrating Equation (3.65). So,

$$\text{stored energy} = \int_0^I LI\,dI$$

$$= \tfrac{1}{2}LI^2 \text{ joule} \tag{3.66}$$

It is interesting to compare this equation with that obtained for the energy stored in a capacitor, Equation (2.51),

$$\text{stored energy} = \tfrac{1}{2}CV^2 \text{ joule}$$

Apart from the obvious difference that inductance replaces capacitance and current replaces voltage, the two expressions are similar. As we have already seen, a current produces a magnetic field, and coils certainly have magnetic fields. However, a voltage produces an electric field and capacitors have electric fields. To take this one stage further, the stored energy per unit volume in a capacitor is given by

energy $= \frac{1}{2} DE$ J m^{-3}

and so we can postulate that the stored energy per unit volume in a coil is

energy $= \frac{1}{2} BH$ J m^{-3} (3.67)

In order to prove this, we can substitute for inductance from Equation (3.46) into Equation (3.66) to give

$$\text{energy} = \tfrac{1}{2} L I^2$$
$$= \tfrac{1}{2} \frac{N^2 \mu A}{l} I^2$$
$$= \tfrac{1}{2} \frac{N^2 I^2}{l^2} \mu \text{ area} \times \text{length}$$
$$= \tfrac{1}{2} H^2 \mu \text{ area} \times \text{length}$$
$$= \tfrac{1}{2} BH \text{ J m}^{-3}$$

So, our supposition was correct. This is an important result in that it shows a duality between electrostatic and electromagnetic fields.

Example 3.11 _____

A 10 mH inductor takes a current of 1.5 A. Determine the energy stored in the magnetic field.

Solution

As the inductance is quoted, we can use Equation (3.66) to give

$$\text{energy} = \tfrac{1}{2} L I^2$$
$$= \tfrac{1}{2} \times 10 \times 10^{-3} \times 1.5^2$$
$$= 11.25 \text{ mJ}$$

There is one major difference between capacitors and inductors: the energy stored in a capacitor stays there for a very long time (it eventually decays away to zero due to leakage) whereas the energy stored in an inductor disappears almost as soon as we remove the current. This is why capacitors are preferred as energy storage devices. (It is true that computer memory boards, and uninterruptable power supplies use capacitors to store energy. However, they can only store enough for very short duration use. If we want a continuous supply, we must use chemical batteries such as lead–acid or nickel–cadmium cells.)

Force between two magnetic surfaces

In the last section we determined the energy stored in the magnetic field of a coil. We were able to draw a comparison with the energy stored in a capacitor to predict the energy density in a magnetic field. Now that we are considering the force between magnetic surfaces, can we use the same procedure?

When we considered electrostatic force, we found that the force between two charged plates is given by (Equation (2.55))

$F = \frac{1}{2} QE$ newton

So, we can postulate that the force between two magnetic surfaces is given by

$F = \frac{1}{2} \phi H$ newton

The method we will use is the same as the one we used to find the electrostatic force. So, let us consider the arrangement shown in Fig. 3.17. This stores a certain amount of energy given by (Equation (3.67))

energy $= \frac{1}{2} BH$ J m^{-3}

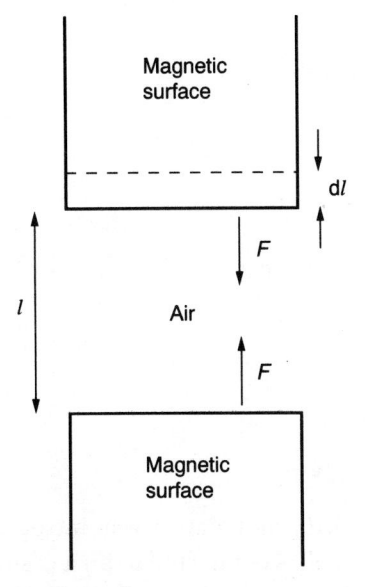

Fig. 3.17 Force between magnetic surfaces

or,

energy $= \frac{1}{2} BH \times$ area $\times l$

Now, there will be a force of attraction between the two surfaces. If we move the top surface by a small amount dl, we do work against the attractive force. This work done must equal the change in stored energy. Thus,

$$F\,dl = \tfrac{1}{2} BH \times \text{area} \times (l + dl) - \tfrac{1}{2} BH \times \text{area} \times l$$
$$= \tfrac{1}{2} BH \times \text{area} \times dl$$

Hence,

$$F = \tfrac{1}{2} BH \times \text{area}$$
$$= \tfrac{1}{2} \phi H \text{ newton} \qquad (3.68)$$

which agrees with our earlier prediction. The flux and field strength in Equation (3.68) relate to the air gap between the two surfaces. So we can write

$$\text{force} = \frac{1}{2} \frac{B^2}{\mu_0} \times \text{area} \qquad (3.69)$$

Example 3.12

The flux density in an air gap between two magnetic surfaces is 5 mWb m^{-2}. Determine the force between the two surfaces if their cross-sectional area is 5 cm^2.

Solution

We can use Equation (3.69) to give

$$\text{force} = \frac{1}{2} \frac{B^2}{\mu_0} \times \text{area}$$

$$= \frac{1}{2} \frac{(5 \times 10^{-3})^2}{4\pi \times 10^{-7}} \times 5 \times 10^{-4}$$

$$= 5 \times 10^{-3} \text{ newton}$$

Low-frequency effects

At the start of Section 3.10, inductance was introduced as a parameter that limits the current when a coil is connected to an a.c. supply. We saw that a back-emf limits the current, with the emf given by (Equation 3.44)

$$e = -L\frac{di}{dt}$$

where the minus sign shows that the induced emf opposes the supply voltage. An alternative way of looking at this is to say that when a coil is connected to an alternating voltage source, an alternating current flows in the coil. We can find the current by solving the following differential equation

$$v_s(t) = L\frac{di}{dt} \qquad (3.70)$$

(We should note that the minus sign is missing because Equation (3.70) uses the supply voltage.)

Figure 3.18(a) shows an inductor connected to an alternating supply. As the source is varying with time, we can write

$i(t) = I_{pk} \sin \omega t$

and so Equation (3.70) becomes

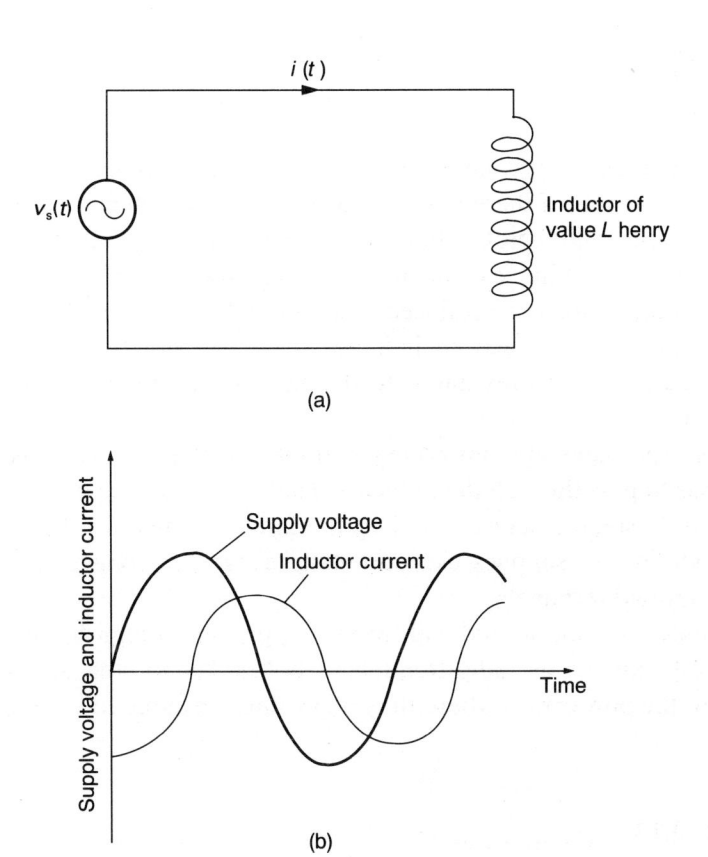

Fig. 3.18 (a) An inductor connected to an a.c. supply; and (b) relationship between inductor current and voltage

$$v_s(t) = L I_{pk} \omega \cos \omega t$$
$$= L I_{pk} \omega \sin(\omega t + 90°) \tag{3.71}$$

So, when connected to an alternating source, the inductor allows a current to flow, with the supply voltage leading the current by 90°. (Figure 3.18(b) shows the relationship between the supply voltage and the inductor current.)

We should note that the back-emf is directly proportional to the angular frequency of the source. We can intuitively reason that the coil current depends on the magnitude of this back-emf: if the frequency is low, the back-emf is small and so the current taken will be high; if the frequency is high, the back-emf is large and so the current taken will be small. We can formalize this by defining the reactance of the inductor, X_L, as

$$X_L = \omega L$$
$$= 2\pi f L \tag{3.72}$$

By combining this result with Equation (3.71) we get, after some rearranging,

$$v_s(t) = i(t) X_L \, \llcorner 90°$$

and so

$$i(t) = \frac{v_s(t)}{X_L} \, \llcorner -90° \tag{3.73}$$

So, when an inductor is connected to an a.c. source, it provides a high resistance path for a.c. signals. If there is also a d.c. voltage, this will allow current to flow. So, we can use an inductor to block a.c. signals but allow d.c. to pass through.

This blocking ability can be used in d.c. power supplies. In such power supplies, a d.c. voltage is produced from a rectified a.c. signal. To remove any a.c. content, we can place an inductor in series with the d.c. output. (The inductor acts as a high resistance to the a.c., but lets the d.c. through without any effect.)

The only problem with this arrangement is that all the current taken from the supply has to pass through the inductor. This current can be quite high (greater than 10 A in some instances) and so the inductor must use heavy-duty wire. This is why power supplies use smoothing capacitors connected between the live and ground terminals.

It is quite common to find coils in the supply rail of high-frequency circuits. These coils, known as radio-frequency chokes, act to stop any signals from getting to the power rails where they might cause trouble to other circuits.

Example 3.13

A 50 mH inductor is connected to a 12 V a.c. supply which has a frequency of 100 Hz. Determine the current taken from the supply.

Solution

The frequency of the supply is 100 Hz, and so the angular frequency is

$$\omega = 2\pi f$$
$$= 2\pi \times 100$$
$$= 200\pi \text{ rad s}^{-1}$$

Now, the reactance of the inductor is given by

$$X_L = \omega L$$
$$= 200\pi \times 50 \times 10^{-3}$$
$$= 31.4 \text{ ohm}$$

Thus the supply current is

$$i_s = \frac{V_s}{X_L}$$
$$= \frac{12}{31.4}$$
$$= 382 \text{ mA}$$

3.12 SOME APPLICATIONS

Towards the end of the last chapter, we considered the motion of electrons through a cathode ray tube. We found that electrons are accelerated by an electron gun at the rear of the CRT. Now, to produce a picture on the phosphor screen at the front of the CRT, the electrons must be deflected by some means. Electrostatic deflection can be used, but this requires very large voltages. Most televisions and monitors use magnetic deflection and focusing coils placed on the neck of the CRT as shown in Fig. 3.19(a). Rather than analyse the operation of deflecting and focusing coils, we will examine the behaviour of an electron beam in a magnetic field. (This is far simpler.)

Figure 3.19(b) shows the situation we are to analyse. As can be seen, we have a stream of electrons travelling in a magnetic field that acts into the page. If this were a current-carrying wire we would have no trouble in finding the force (Equation (3.9) and Section 3.7 apply). As a wire is simply a conduit for the passage of charge, we can consider the electron flow as a current, and use

$$F = BIl$$

to find the force on the electron stream.

Now, current is equal to the rate of change of charge, and so

112 *Electromagnetic fields*

(a)

(b)

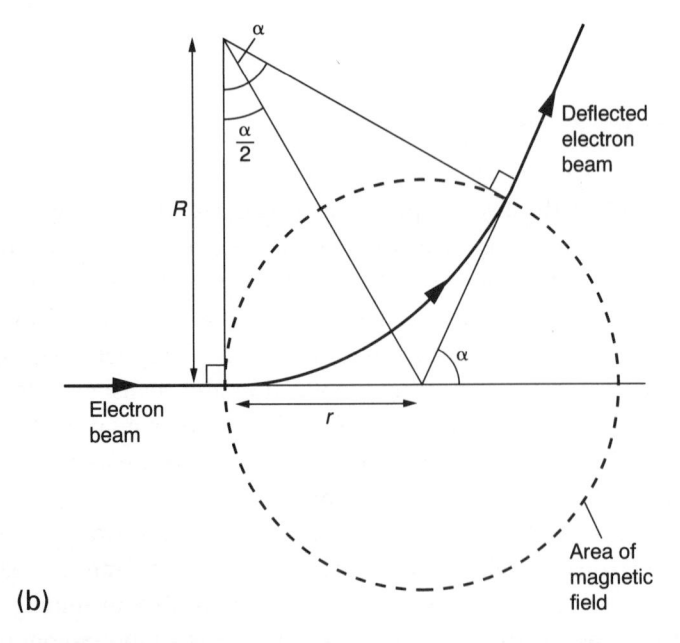

Fig. 3.19 (a) A CRT with electromagnetic focusing coils; and (b) deflection of an electron beam by a magnetic field

3.12 Some applications

$$F = B \frac{dQ}{dt} l$$

Let us assume that N electrons pass through the field in time t and that, during this time, they travel the distance l. So, we can write

$$F = B N q \frac{l}{t}$$

$$= B N q v$$

where v is the velocity of the electrons. Thus the force on a single electron is

$$F = B q v \qquad (3.74)$$

We should note that velocity in Equation (3.74) is the component at right angles to the magnetic field. As the electron gets deflected, this velocity component tends to move the electron in a circular path (the dotted line in Fig. 3.19(b)).

As the electron moves in a circular path, it will experience a centripetal force, acting in opposition to the magnetic force, of

$$F = \frac{mv^2}{R} \qquad (3.75)$$

The motion of the electron is governed by Equations (3.74) and (3.75) as the following example shows.

As an example, let us consider a magnetic field, acting into the page, that is only present in a circle of radius r. As Fig. 3.19(b) shows, an electron entering this field is deflected around an arc of a circle of radius R. Now, we wish to find the angle α through which the magnetic field deflects the electron.

Let us take a B field of 0.01 Wb m^{-2} present over a circle of radius 1.5 cm. If we assume that the electron was accelerated by a potential difference of 30 kV, its velocity will be 1.0×10^8 m s^{-1}. So, it will experience a force of (Equation (3.74))

$$F = 0.01 \times 1.6 \times 10^{-19} \times 1.0 \times 10^8$$
$$= 1.6 \times 10^{-13} \text{ N}$$

The centripetal force on the electron will be (Equation (3.75))

$$F = \frac{9.1 \times 10^{-31} \times (1.0 \times 10^8)^2}{R}$$

$$= \frac{9.1 \times 10^{-15}}{R} \text{ N}$$

So, by equating these forces we get

$$R = \frac{9.1 \times 10^{-15}}{1.6 \times 10^{-13}}$$

$$= 5.7 \text{ cm}$$

Now, a bit of simple geometry – see Fig. 3.19(b) – shows that

$$\tan \frac{\alpha}{2} = \frac{r}{R}$$

$$= \frac{1.5}{5.7}$$

$$= 0.26$$

This gives

$$\frac{\alpha}{2} = 14.7°$$

and so the deflection is 29.5°.

As a matter of interest, if the B field occupies a larger area, the electron path will be circular. This property is put to good use in particle accelerators and tokamak fusion reactors.

Figure 3.20(a) shows a photograph of the Joint European Torus, or JET, tokamak fusion reactor at Culham, Oxfordshire. This is an experimental machine in which large amounts of energy can be liberated by the fusion of the hydrogen isotopes deuterium and tritium. (This is the same process that goes on in the Sun.) To initiate fusion, it is necessary to control a plasma with a temperature in excess of 200 million degrees Celsius! As the plasma consists of positively charged nuclei, an obvious way of confining it is to use a magnetic field.

As can be seen in Fig. 3.20(b), the reactor vessel is toroidal in shape with an inner radius of 1.25 m and an outer radius of 3 m. This vessel is surrounded by 32 electromagnetic coils arranged around the outside. Each coil consists of 24 turns of wire carrying a current of 67 kA, which results in a maximum field of 3.4 Wb m^{-2} at the centre of the vessel. Eight large iron-cored transformers (four of which can be seen in Fig. 3.20(a)) generate an additional field that serves to confine, as well as heat, the plasma. (In effect, the plasma forms a single-turn secondary, with energy being supplied from the primary coil to the plasma via the iron limbs.) This arrangement has resulted in plasma currents of up to 7 MA, and this serves to heat the plasma to about 50 million °C. Further heating is provided by 20 MW of neutral particle injection (approximately 140 keV) and 24 MW of radio frequency power (23–57 MHz).

Although JET is not designed to break even, i.e. generate more energy than it consumes, the project has produced 1.7 MW of energy. The conclusion is that a larger version of the machine could generate useful power. In view of this, it is proposed to construct a new torus with an inner radius of 3 metre, and an outer radius of 8 metre. If all goes according to plan, the new machine, known as the International Thermonuclear Experimental Reactor, or ITER, will begin operations in 2008. The design of this machine calls for a magnetic field of 13 Wb m^{-2} to be generated by superconducting coils carrying 40–60 kA of current, resulting in a stored energy of about 100 GJ! This is clearly engineering on a grand scale.

3.13 SUMMARY

We started this chapter by examining Coulomb's law as applied to magnetostatics. This introduced us to the general ideas of magnetic flux, magnetic flux density, and magnetic field strength. This initial study was only concerned with isolated magnetic monopoles.

We then went on to consider the magnetic field produced by a current-carrying wire – electromagnetism. This introduced us to the Biot–Savart law which is a fundamental law of electromagnetism. We then went on to review our ideas of magnetic flux density and magnetic field strength. The relevant formulae are summarized here:

$$dH = \frac{Idl}{4\pi r^2} \sin\theta \tag{3.76}$$

$$dB = \mu \frac{Idl}{4\pi r^2} \sin\theta \, p_N \tag{3.77}$$

$$dF = p \, dB \tag{3.78}$$

We then examined the magnetic field produced by a variety of current-carrying wires: a long straight conductor; a circular conductor; and a solenoid, or coil. As well as introducing us to applications of the Biot–Savart law, these examples also introduced us to Ampère's law. We then met magnetic potential. This was defined as the work done in moving a unit pole around a magnetic field, and introduced us to the idea of reluctance as the resistance to the flow of flux.

In Section 3.11 we were introduced to the idea of inductance. We determined the inductance of various arrangements: simple coil; an isolated wire; coaxial cable; twin feeder; and microstrip lines. The inductances are reproduced here:

(a)

Fig. 3.20 (a) The Joint European Torus, JET, fusion project

Fig. 3.20 (b) Cut-away of the JET reactor vessel

simple coil $$L = \frac{N^2 \mu A}{l} \text{ H} \tag{3.79}$$

isolated wire $$L' = \frac{\mu_0}{8\pi} \text{ H m}^{-1} \text{ (internal)} \tag{3.80}$$

coaxial cable $$L' = \frac{\mu_0}{8\pi} + \frac{\mu_0}{2\pi} \ln(b/a) \text{ H m}^{-1} \tag{3.81}$$

twin feeder $$L' = \frac{\mu_0}{4\pi} + \frac{\mu_0}{\pi} \ln\left(\frac{d-a}{a}\right) \text{ H m}^{-1} \tag{3.82}$$

microstrip $$L' = \frac{\mu_0}{8\pi} + \mu_0 \frac{h}{w} \text{ H m}^{-1} \tag{3.83}$$

or, $$L' = \frac{\mu_0}{8\pi} + \frac{\mu_0}{2\pi} \ln(h/w) \text{ H m}^{-1} \tag{3.84}$$

We also came across a very important relationship: the capacitance and inductance per unit length are related by

$$L'C' = \mu\epsilon \tag{3.85}$$

Thus we can find the inductance if we know the capacitance.

In common with capacitors, we found that inductors can also store energy. However, there is one major difference: a capacitor can store energy for a very long time, whereas the energy stored in a magnetic field disappears when the field collapses. We found that the stored energy can be given by

$$\text{energy} = \tfrac{1}{2}LI^2 \text{ J} \tag{3.86}$$

or,

$$\text{energy} = \tfrac{1}{2}BH \text{ J m}^{-3} \tag{3.87}$$

We were then introduced to the reactance of an inductor given by

$$X_\text{L} = 2\pi f L \tag{3.88}$$

with the supply voltage leading the coil current by 90°.

As with the last chapter, we could have examined parallel and series combinations of inductors. The reason why we didn't do this is that it is very easy to wind a coil on a former. Thus non-standard value inductors are far simpler to produce than non-standard capacitors.

We concluded the chapter with a brief examination of electromagnetic deflection, and plasma confinement in a fusion reactor. In the first example, we saw that a quite moderate field of $0.01\,\text{Wb}\,\text{m}^{-2}$ can produce an electron deflection of 30°. By way of contrast, the magnetic field in use at the Joint European Torus fusion reactor is some 340 times as large. This is needed to confine a 200 million degree Celsius plasma.

4 Electroconductive fields

We now come to the last field system we are going to study – electroconductive fields. As electrical engineers, this is probably the most familiar field system: every time we build an electrical circuit, we use electroconduction – charge flow around the circuit causes a current. However, what exactly is the nature of the current, how is it conducted, and why does it flow? We can answer these questions by studying electroconductive fields.

4.1 CURRENT FLOW

When we considered electrostatics in Chapter 2, we saw that like charges repel each other. Electrical conductors put this effect to good use. To be a good conductor of electricity, the atoms in the material must have a large number of electrons in the conduction band. Most metals satisfy this requirement, and are malleable enough to be drawn into wires.

Let us consider what happens when we connect a wire to a source of electrons, a battery for instance (see Fig. 4.1). The battery produces electrons at the negative terminal by chemical means. So, when the negative terminal produces an electron, it repels electrons in the wire, so forcing an electron out

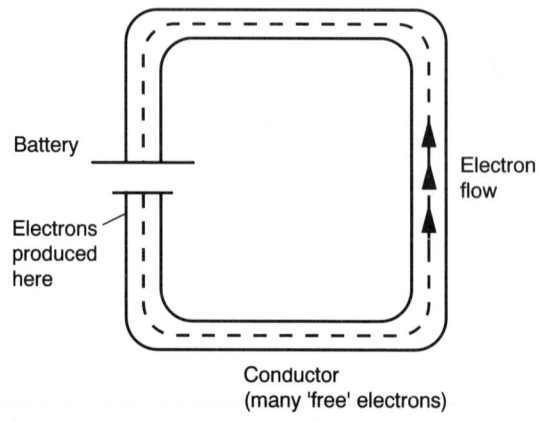

Fig. 4.1 A conductor connected to a voltage source

4.2 Potential and electric field strength

of the wire into the positive battery terminal. We define the current through the wire as the rate of flow of charge, i.e.

$$i = \frac{dQ}{dt} \tag{4.1}$$

with units of coulomb per second, or ampere.

So, current is equal to the rate of flow of charge. This raises the question: what forces the charge around the circuit? This is where we reintroduce the ideas of potential and electric field strength.

4.2 POTENTIAL AND ELECTRIC FIELD STRENGTH

In Chapter 2 we came across potential as a measure of the work done against an electrostatic field. Now that we are considering an electroconductive field, what part does potential play?

Figure 4.2 shows a piece of conducting material connected across a battery. Now, the positive terminal of the battery will attract electrons, while the negative terminal will repel electrons. So, if we move a positive test charge from the negative terminal to the positive terminal, we have to do work against a field. The precise amount of work done is equal to the potential of the battery. (This is precisely the same definition of potential that we used when we considered electrostatics.)

As Fig. 4.2(b) shows, we can draw lines of equal potential in the conductor.

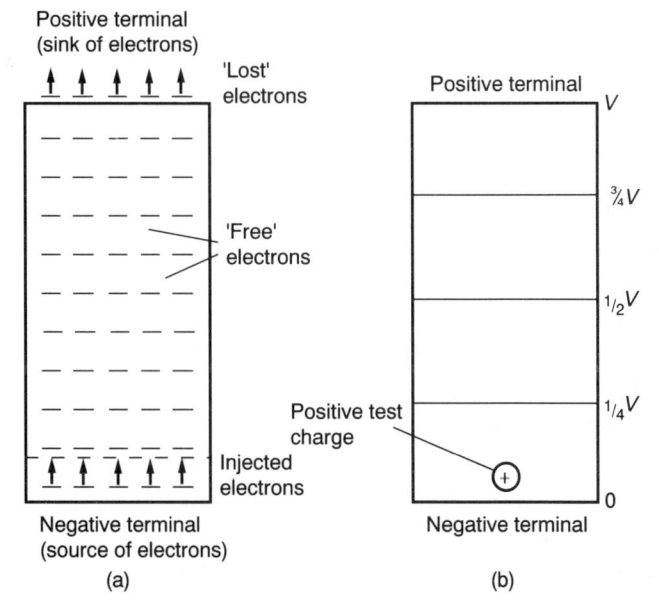

Fig. 4.2 (a) Electron flow in a conductor connected to a voltage source; and (b) equipotentials in a conductor

122 *Electroconductive fields*

This diagram shows that the potential increases as we move towards the positive terminal, confirming that we must do work against the field as we move our positive test charge along the conductor.

Let us now consider what happens to an electron injected into the conductor from the negative terminal of the battery. The positive terminal attracts this electron, while the negative terminal repels it. Thus the electron experiences an electrostatic force that moves it up the conductor. The magnitude of this E field is given by

$$E = \frac{V}{l} \tag{4.2}$$

where l is the length of the conductor.

So, we have seen that a voltage source sets up an electric field in any conductor connected across it. We have also seen that the potential of the source is equal to the work done in moving a positive test charge around the external circuit. Charge flowing around the external circuit does so under the influence of the external field. The current represents this charge flow, and this is the subject of the next section.

4.3 CURRENT DENSITY AND CONDUCTIVITY

We have just seen that charges move around a circuit under the influence of an electric field. This charge flow can be regarded as a current by virtue of (Equation (4.1)),

$$i = \frac{dQ}{dt}$$

What precisely happens when the source injects an electron into the conductor? Well, this electron immediately enters an electric field that accelerates it towards the positive terminal. However, because the electron is travelling in a crystal lattice, it cannot travel very far before it meets an atom. Metal atoms have a large number of electrons available for conduction, and so, when this electron meets a metal atom, the injected electron will dislodge a conduction-band electron from the atom. The injected electron will then be caught by the metal atom, and the dislodged electron will travel through the lattice under the influence of the electric field. Of course this new electron will soon meet a metal atom, and the whole process repeats along the length of the conductor. Thus the original electron takes a long time to reach the end of the wire, but the effect is felt after quite a short time. (In fact the disturbance travels at the speed of light in the material surrounding the conductor.)

So, the original charge carriers from the battery cannot travel very far through the lattice before meeting a metal atom. Let us consider the motion of an electron across a volume of conductor (see Fig. 4.3). These electrons will

4.3 Current density and conductivity

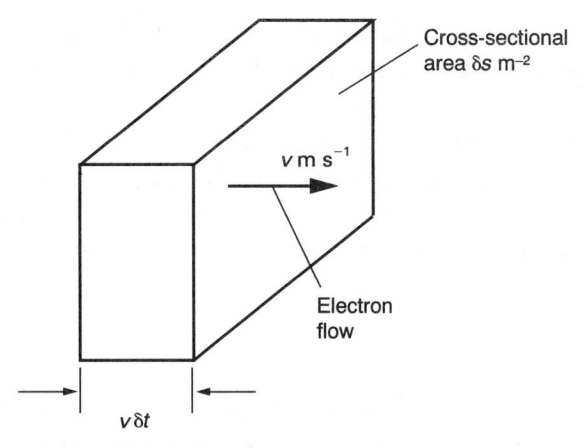

Fig. 4.3 Elemental section through a current-carrying conductor

have an average velocity, known as the drift velocity, with symbol v m s^{-1}. Thus, in time δt, the electrons will have moved a distance $v\delta t$ metre. If N is the number of electrons per unit volume, m^{-3}, and δs is the cross-sectional area of the volume, m^{-2}, the amount of charge passing through the area δs is

$$\delta Q = Nq\, v\delta t\, \delta s \text{ coulomb} \tag{4.3}$$

As current is the rate of flow of charge, the current through this area is

$$i = \frac{\delta Q}{\delta t}$$

$$= Nqv\, \delta s \text{ ampere} \tag{4.4}$$

Thus the current per unit area, or the current density, is

$$J = \frac{i}{\delta s}$$

$$= Nqv \text{ A m}^{-2} \tag{4.5}$$

We have reasoned that charges move through the conductor under the influence of an electric field. Thus we can intuitively reason that the velocity of the electrons will depend on the value of E. Specifically, we can write,

$$v = \mu_e E \text{ m s}^{-1} \tag{4.6}$$

where μ_e is the mobility of the electrons in the lattice. By combining Equations (4.5) and (4.6) we get

$$J = Nq\mu_e E$$

or,

$$J = \sigma E \tag{4.7}$$

where σ is known as the conductivity of the material.

So, a material constant relates the current density to the electric field strength. This is identical in form to the corresponding relationships in electrostatics, $D = \epsilon E$, and electromagnetism, $B = \mu H$.

According to Equation (4.7) the current density is directly dependent on the electric field strength. Thus if we double the electric field strength, the current density will double. Of course, if the conductivity rises, the current density will also rise for a given E field. Thus we can see that good conductors must have a large conductivity.

Example 4.1

A 12 volt potential is set up across a conductor of length 50 cm. The conductor is made of copper, $\sigma = 58$ MS m^{-1}, and has a uniform cross-sectional area of 10 cm^2. Determine the electric field strength, the current density, and the current that flows through the conductor.

Solution

The potential across the conductor is 12 volt, and the length is 50 cm. So, the electric field strength is

$$E = \frac{12}{50 \times 10^{-2}}$$
$$= 24 \text{ V m}^{-1}$$

Now,

$$J = \sigma E$$

and so,

$$J = 58 \times 10^6 \times 24$$
$$= 1.4 \times 10^9 \text{ A m}^{-2}$$

As the cross-sectional area of the sample is 10 cm^2, the current is

$$I = 1.4 \times 10^9 \times 10 \times 10^{-4}$$
$$= 1.4 \text{ MA}$$

It is a good job that this is simply an example, because such a large current would quickly vaporize the conductor!

4.4 RESISTORS

In the last section we saw that the current density and electric field strength are related by (Equation (4.7))

$$J = \sigma E$$

Now, by substituting for J and E we get

$$\frac{I}{A} = \sigma \frac{V}{l}$$

which, after some rearranging, gives

$$\frac{V}{I} = \frac{l}{\sigma A}$$

or

$$\frac{V}{I} = R \tag{4.8}$$

where R is the resistance of the conductor in ohms, given by

$$R = \frac{l}{\sigma A} \tag{4.9}$$

(An alternative expression for the resistance is

$$R = \frac{\rho l}{A} \tag{4.10}$$

where ρ is the resistivity of the material ($\rho = 1/\sigma$). In this book, however, we will use conductivity rather than resistivity.)

Equation (4.8) is known as Ohm's law, named after Georg Simon Ohm, 1789–1854, the German physicist who studied electroconductive fields. Stated simply it says that the current flowing through a conductor is directly proportional to the voltage across it, and inversely proportional to the resistance. So, the lower the resistance, the higher the current.

In passing through any resistor, the current has to do work, and this generates heat. (A familiar example of this is the heat generated by an electric fire.) The amount of power dissipated by a resistor is given by

$$\text{power} = I^2 R \text{ watts} \tag{4.11}$$

Example 4.2

A 12 volt potential is set up across a conductor of length 50 cm. The conductor is made of copper, $\sigma = 58 \text{ MS m}^{-1}$, and has a uniform cross-sectional area of 10 cm². Determine the resistance of the conductor, and calculate the heat generated.

Solution

The conductor has a regular cross-sectional area, and so the resistance is given by

$$R = \frac{l}{\sigma \times \text{area}}$$

$$= \frac{50 \times 10^{-2}}{58 \times 10^6 \times 10 \times 10^{-4}}$$

$$= 8.6 \ \mu\Omega$$

The power dissipated by this resistance is

$$\text{power} = I^2 R$$

$$= \left(\frac{12}{8.6 \times 10^{-6}}\right)^2 8.6 \times 10^{-6}$$

$$= 16.7 \ \text{MW}$$

So, this resistor is a very efficient producer of heat. Again, we must use care here because the heat generated is so high that it might vaporize the conductor.

Before we consider the resistance of the various transmission lines we have covered in previous chapters, let us examine the resistance of a capacitor.

Capacitors

At first sight it might seem that capacitors are out of place in a chapter on electroconduction. However, we should remember that capacitors have a dielectric between the two plates, and this material can conduct a small amount of current – the leakage current.

If the capacitor has a regular cross-section, we can find the resistance from

$$R = \frac{d}{\sigma A} \tag{4.12}$$

with the capacitance given by

$$C = \frac{\epsilon A}{d} \tag{4.13}$$

So, a non-ideal capacitor can have both displacement current and conduction current between the two plates. The proportion of conduction current to displacement current is a measure of the loss of the capacitor.

Figure 4.4 shows the two currents drawn on an Argand diagram. As this figure shows, the conduction current is drawn on the real axis, and the displacement current is drawn on the negative imaginary axis. (It is drawn on the negative axis because it leads the conduction current by 90°.) Now, the conduction current is given by

$$I_{con} = \frac{V}{R}$$

$$= \frac{V \times \sigma \times \text{area}}{d} \tag{4.14}$$

and the displacement current is given by

$$I_{disp} = \frac{V}{X_C}$$

$$= \frac{V}{1/\omega C}$$

$$= V\omega \frac{\epsilon \times \text{area}}{d} \tag{4.15}$$

As Fig. 4.4 shows, the ratio of these two currents is also equal to the tangent of the angle δ. Thus,

$$\tan \delta = \frac{I_{con}}{I_{disp}}$$

$$= \frac{V \times \sigma \times \text{area}}{d} \times \frac{d}{V\omega\epsilon \times \text{area}}$$

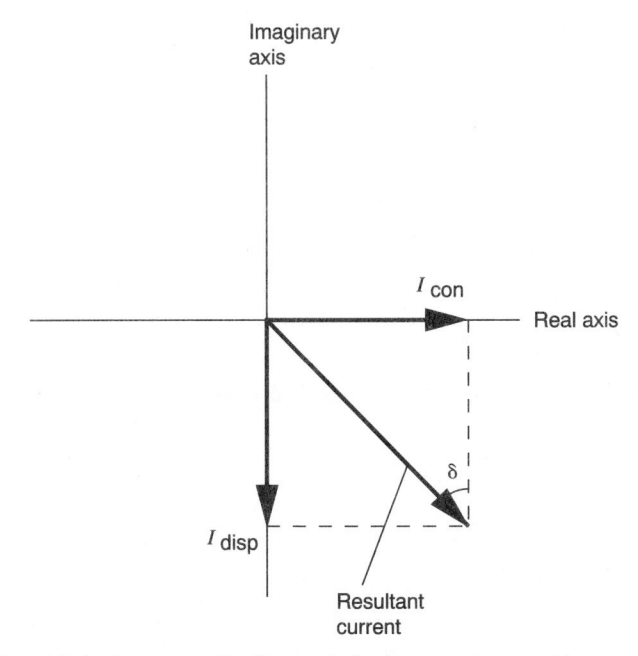

Fig. 4.4 Relationship between conduction and displacement current in a non-ideal capacitor

i.e.

$$\tan \delta = \frac{\sigma}{\omega \epsilon} \tag{4.16}$$

Equation (4.16) shows that $\tan \delta$ is a measure of the loss in the capacitor. Thus $\tan \delta$ is known as the loss tangent of the capacitor. Although readers may wonder at the use of such a quantity, it is quite important as the following example shows.

Example 4.3

A parallel plate capacitor consists of two metal plates, of area 2 cm², separated by 3 μm of porcelain with $\epsilon_r = 5.7$ and $\sigma = 2 \times 10^{-13}$ S m⁻¹. The capacitor is connected to a 12 volt 50 Hz supply. Determine the conduction current, and compare it to the displacement current. In addition, calculate the loss tangent at frequencies of 50 Hz, 1 MHz, and 100 MHz.

Solution

The capacitor is connected to a 12 volt, 50 Hz supply. As the dielectric is lossy, the capacitor will also have some resistance given by

$$R = \frac{d}{\sigma \times \text{area}}$$

$$= \frac{3 \times 10^{-6}}{2 \times 10^{-13} \times 2 \times 10^{-4}}$$

$$= 7.5 \times 10^{10} \, \Omega$$

By applying Ohm's law, the leakage current is

$$I = \frac{12}{7.5 \times 10^{10}}$$

$$= 0.16 \text{ nA}$$

To find the displacement current, we need to find the capacitance. So,

$$C = \frac{\epsilon_0 \epsilon_r \times \text{area}}{d}$$

$$= \frac{8.854 \times 10^{-12} \times 5.7 \times 2 \times 10^{-4}}{3 \times 10^{-6}}$$

$$= 3.4 \text{ nF}$$

At 50 Hz, the reactance is

$$X_C = \frac{1}{2\pi f C}$$

$$= \frac{1}{2\pi \times 50 \times 3.4 \times 10^{-9}}$$

$$= 946 \text{ k}\Omega$$

and so the displacement current is

$$I = \frac{12}{946 \times 10^3}$$

$$= 12.7 \text{ μA}$$

Thus the displacement current is 79 000 times greater than the leakage current. This difference is quite fortunate, otherwise capacitors would be useless!

The loss tangent is given by (Equation (4.16))

$$\tan \delta = \frac{\sigma}{\omega \epsilon}$$

and so

$$\tan \delta = \frac{2 \times 10^{-13}}{2\pi f \times 8.854 \times 10^{-12} \times 5.7}$$

$$= \frac{6.3 \times 10^{-4}}{f}$$

The following table compares the loss tangent at various frequencies.

Frequency (Hz)	tan δ
50	1.3×10^{-5}
1×10^6	6.3×10^{-10}
100×10^6	6.3×10^{-12}

Note that tan δ is increasing with frequency. This is because the reactance of the capacitor is decreasing, so increasing the displacement current. The leakage current is relatively independent of frequency and so the lower the value of tan δ, the greater the insulating properties of the material. The converse is also true – the greater the loss tangent, the better the conduction properties of the material.

Before we leave the parallel plate capacitor, let us return to our expressions for capacitance and resistance, Equations (4.12) and (4.13). If we multiply these two together, we get

$$RC = \frac{d}{\sigma A} \frac{\epsilon A}{d}$$

$$= \frac{\epsilon}{\sigma} \qquad (4.17)$$

Thus it would appear that we have another fundamental relationship linking resistance and capacitance. However, this is only one example, and we should not start applying this 'rule' with enthusiasm just yet.

Coaxial cable

With coaxial cable we have two resistances to consider: the resistance of the inner conductor; and the resistance between the inner and outer conductors. As regards the first resistance, the cross-section of the inner conductor is circular, and so the resistance is given by

$$R = \frac{l}{\sigma A} \, \Omega$$

or,

$$R' = \frac{1}{\sigma A} \, \Omega \, \text{m}^{-1} \qquad (4.18)$$

Let us now turn to the resistance between the inner and outer conductors. If the inner conductor is at a potential of V volt above the outer conductor, there will be an electric field set up in the dielectric. In addition, leakage current will flow in a radial direction outwards from the inner conductor (see Fig. 4.5). Now, the current density at a certain radius will be

$$J = \frac{I}{2\pi r \times \text{length}} \qquad (4.19)$$

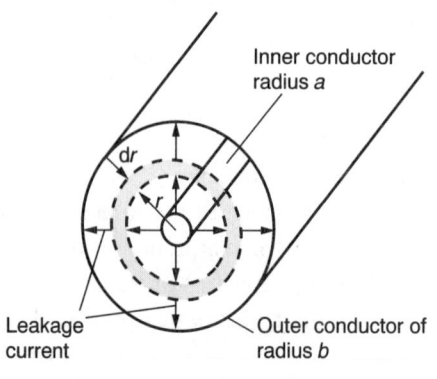

Fig. 4.5 Leakage current in a length of coaxial cable

Thus the E field at this radius is

$$E = \frac{J}{\sigma}$$

$$= \frac{I}{\sigma 2\pi r \times \text{length}} \qquad (4.20)$$

Let us consider a thin, hollow tube of radius r and thickness dr. The potential across this tube, dV, is

$$dV = \frac{-I}{\sigma 2\pi r \times \text{length}} dr$$

and so the total voltage between the inner and outer conductors is

$$\int_0^V dV = \int_b^a \frac{-I}{\sigma 2\pi r \times \text{length}} dr$$

Hence,

$$V = \frac{-I}{2\pi\sigma \times \text{length}} \int_b^a \frac{1}{r} dr$$

$$= \frac{I}{2\pi\sigma \times \text{length}} \left| \ln r \right|_a^b$$

$$= \frac{I}{2\pi\sigma \times \text{length}} \ln(b/a)$$

Thus the leakage resistance of the cable is

$$R = \frac{1}{2\pi\sigma \times \text{length}} \ln(b/a) \; \Omega \qquad (4.21)$$

or,

$$R' = \frac{\ln(b/a)}{2\pi\sigma} \; \Omega \, \text{m} \qquad (4.22)$$

As we have already seen, the capacitance of such a cable is given by (Equation (2.37))

$$C' = \frac{2\pi\epsilon_0\epsilon_r}{\ln(b/a)} \; \text{F m}^{-1}$$

So, the product of R' and C' is

$$R'C' = \frac{\ln(b/a)}{2\pi\sigma} \frac{2\pi\epsilon_0\epsilon_r}{\ln(b/a)}$$

$$= \frac{\epsilon_0\epsilon_r}{\sigma} \qquad (4.23)$$

132   Electroconductive fields

We obtained the same result when we multiplied the resistance and capacitance of a parallel plate capacitor together. So we appear to have a new rule that enables us to find the resistance of a field system if we know the capacitance.

Example 4.4

A 500 m length of coaxial cable has an inner conductor of radius 2 mm, and an outer conductor of radius 1 cm. The conductivity is 3×10^{-4} S m^{-1}, while the conductivity of the inner core is 58 MS m^{-1}. Determine the series and shunt resistance of the cable.

Solution

The cable is 500 metre long, and so we can use Equation (4.20) to find the shunt resistance. Thus,

$$R = \frac{\ln(b/a)}{2\pi\sigma \times \text{length}}$$

$$= \frac{\ln(1 \times 10^{-2}/2 \times 10^{-3})}{2\pi \times 3 \times 10^{-4} \times 500}$$

$$= \frac{\ln 5}{0.94}$$

$$= 1.7 \, \Omega$$

We could have obtained this result from a knowledge of the cable capacitance. This cable is the same as that used in Example 2.9 in Section 2.9. From this example, the cable capacitance is 86.4 nF, and so the shunt resistance is

$$R = \frac{\epsilon_0 \epsilon_r}{\sigma C}$$

$$= \frac{8.854 \times 10^{-12} \times 5}{3 \times 10^{-4} \times 86.4 \times 10^{-9}}$$

$$= 1.7 \, \Omega$$

As the inner conductor has a regular cross-sectional area, we can use Equation (4.9) to give

$$R = \frac{l}{\sigma A}$$

$$= \frac{500}{58 \times 10^6 \times \pi \times (2 \times 10^{-3})^2}$$

$$= 0.8 \, \Omega$$

So, the cable has a series resistance of 0.8 Ω, and a shunt resistance of 1.7 Ω. We have already seen that this cable has inductance and capacitance as well, and we must take these into account when determining whether the cable is useful for a particular purpose.

Twin feeder

Let us now consider twin feeder. This type of cable is usually air spaced and so the shunt resistance is zero. (If air conducted electricity, our electricity bills would be even higher than now!) So, air-spaced twin feeder only has the series resistance of the conductors. Thus,

$$R = \frac{l}{\sigma A} \, \Omega \tag{4.24}$$

or,

$$R' = \frac{1}{\sigma A} \, \Omega \, \text{m}^{-1} \tag{4.25}$$

Of course, if the twin feeder were not air spaced, but surrounded by a lossy dielectric, the shunt resistance would be given by

$$R = \frac{\ln((d-a)/a)}{\pi \sigma \times \text{length}} \, \Omega \tag{4.26}$$

where we have made use of our new relationship, (Equation (4.23).

Microstrip lines

We seldom need to consider the shunt resistance of a length of microstrip. This is because printed circuit boards are usually made of PTFE or Teflon, $\sigma \approx 10^{-16}$ S m^{-1}, and this makes them very poor conductors. Instead, the voltage drop along a length of track is very important.

The metallization on printed circuit boards usually has a uniform cross-sectional area, and so the resistance of a length of track is given by

$$R = \frac{l}{\sigma A} \, \Omega \tag{4.27}$$

where A is the cross-sectional area of the track. As copper is usually used on pcbs, the resistance per cm is

$$R' = \frac{1}{58 \times 10^8 \times w \times t}$$

$$= \frac{1.7 \times 10^{-10}}{w \times t} \, \Omega \, \text{cm}^{-1} \tag{4.28}$$

where w is the width of the track, and t is the thickness of the metallization.

134 *Electroconductive fields*

Example 4.5

A printed circuit board has a 10 cm long, 5 volt supply rail etched on to it. This track feeds two logic devices, one of which is half-way along the line while the other is at the end of the track. If the first device takes a current of 300 mA while the second only takes 10 mA, determine the supply voltage at both devices. If the second device suddenly takes 300 mA, determine the new supply voltages. Assume that the copper metallization is 20 μm thick, and that the track width is 2 mm.

Solution

We have a 10 cm length of track feeding 5 volts to two logic circuits. The first is half-way along the track, and takes a current of 300 mA. As the track has a uniform cross-sectional area, and it is made of copper, the resistance per cm is given by (Equation (4.28))

$$R' = \frac{1.7 \times 10^{-10}}{w \times t} \, \Omega \, \text{cm}^{-1}$$

$$= \frac{1.7 \times 10^{-10}}{2 \times 10^{-3} \times 20 \times 10^{-6}} \, \Omega \, \text{cm}^{-1}$$

$$= 4.25 \, \text{m}\Omega \, \text{cm}^{-1}$$

Now, the first device is 5 cm from the source. So, the resistance of the first length is

$$R = 4.25 \times 10^{-3} \times 5$$
$$= 21.25 \, \text{m}\Omega$$

The current flowing through this track is the sum of the individual currents, i.e.

$$I = 300 + 10 \, \text{mA}$$
$$= 310 \, \text{mA}$$

So, the voltage drop is

$$V = IR$$
$$= 310 \times 10^{-3} \times 21.25 \times 10^{-3}$$
$$= 6.6 \, \text{mV} \quad \text{over the first section}$$

Thus the first device has a supply voltage of

$$V_1 = 5 - 6.6 \times 10^{-3}$$
$$= 4.9934 \, \text{V}$$

The current through the second section of the line is 10 mA. Thus the voltage drop across this section is

$$V = IR$$
$$= 10 \times 10^{-3} \times 21.25 \times 10^{-3}$$
$$= 0.213 \text{ mV}$$

and so the second device has a supply voltage of

$$V_2 = 4.9934 - 0.213 \times 10^{-3}$$
$$= 4.9932 \text{ V}$$

By following a similar procedure, the supply voltages when the second device takes 300 mA are

$$V_1 = 4.98725 \text{ V}$$

and

$$V_2 = 4.98088 \text{ V}$$

Although the difference in supply rails is small in this example, it does show the dangers of making the supply rail too thin. We can also see the effect of switching on a logic gate: the power rail will drop for all devices supplied by the line. If the logic devices are turned on and off, as in a sequential logic design, this can lead to sudden voltage variations throughout the board. The solution is to use decoupling capacitors and wide tracks. (Decoupling capacitors are placed between the supply and ground very close to the logic devices. As the capacitors are charged to the supply rail, they can supply any sudden demand for current. Thus the voltage along the power rail will not suddenly drop.)

Kirchhoff's voltage and current laws

While we are considering electroconductive fields, let us take some time to examine Kirchhoff's laws which are more often treated as 'circuit' laws. Figure 4.6(a) shows a circuit in which two resistors are connected in series. This combination is connected to a source of potential V. Now, what happens as we

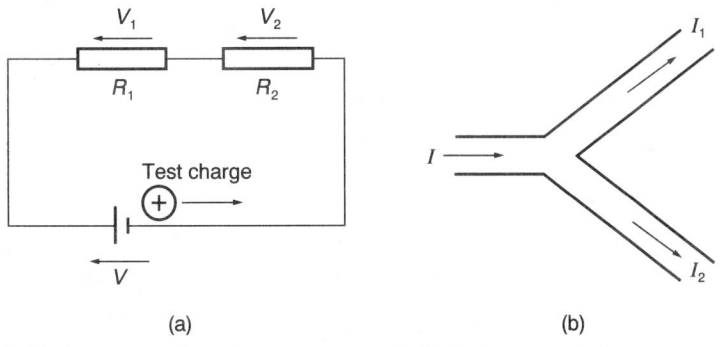

Fig. 4.6 (a) Series connection of two resistors; and (b) current split between two conductors

move a positive test charge from the negative terminal to the positive terminal via the external circuit?

Let us move the test charge from the negative terminal through R_2 to the junction between the two resistors. As there is an E field set up in resistor R_2, we have to do work against the field. The magnitude of the E field can be written as

$$E_2 = \frac{V_2}{l_2} \tag{4.29}$$

where l_2 is the length of resistor R_2. Now, the force on our unit test charge is

$$F_2 = \frac{V_2}{l_2} \times 1$$

$$= \frac{V_2}{l_2} \text{ newton}$$

and so the work done in moving through the resistor is

$$\text{work done} = \frac{V_2}{l_2} \times l_2$$

$$= V_2 \tag{4.30}$$

Similarly, the work done in moving through resistor R_1 is

$$\text{work done} = \frac{V_1}{l_1} \times l_1$$

$$= V_1 \tag{4.31}$$

Hence the total work done in moving the test charge around the external circuit is

$$\text{total work done} = V_1 + V_2 \tag{4.32}$$

This total work done must be equal to the potential of the source. (Any shortfall in the potential would be dropped across another resistance, and so we would have to do additional work against another field.) Thus,

$$V = V_1 + V_2 \tag{4.33}$$

i.e. the sum of the potential drops around an external circuit is equal to the supply potential. This is Kirchhoff's voltage law. So, the potential differences around a circuit must add up to the supply potential. However, what happens to the current at a junction in an external circuit?

Figure 4.6(b) shows a current-carrying conductor splitting into two paths. Let us consider a current flowing through the left-hand conductor. This current will split between the two right-hand conductors, with the resistance of the two branches determining the exact proportions of the split. Now, the only current entering the junction, I, is from the left-hand conductor, and the only currents leaving the junction, I_1 and I_2, do so through the right-hand conductors. As there are no other sources of charge and no sinks of charge, we can reason that the current entering the junction is equal to the current leaving. Thus,

$$I = I_1 + I_2 \qquad (4.34)$$

In general we can state that the current entering a junction must equal the current leaving. This is Kirchhoff's current law which we can regard as an example of 'good housekeeping' – we have to account for all sources and sinks of current. A similar situation exists in all the field problems we have met – the flux entering a surface must equal the flux leaving the surface if the surface does not enclose a source of flux.

Combinations of resistors

Like capacitors, resistors come in standard values. Sometimes this can be very inconvenient, and so we have to make our own values. One way to do this is to trim an existing resistor by gently filing away at the body! This requires a very steady hand, and a great deal of patience. An alternative is to combine resistors in series or parallel.

Figure 4.7(a) shows two resistors, R_1 and R_2, in parallel. We require to find the equivalent resistance of this arrangement. Let us connect a d.c. source, V_s, to the resistors. Now, both resistors have the same voltage across them, but they will pass different currents. So,

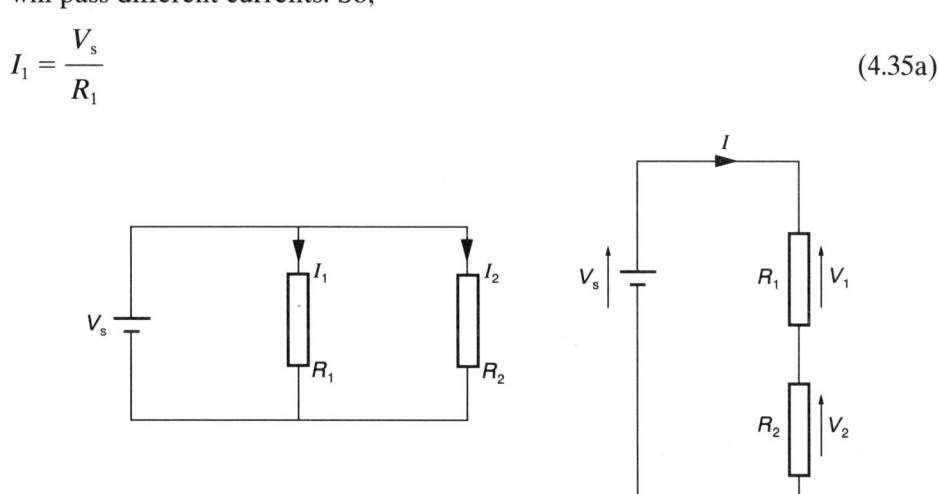

$$I_1 = \frac{V_s}{R_1} \qquad (4.35a)$$

Fig. 4.7 (a) Parallel combination of two resistors; and (b) series combination of two resistors

and,

$$I_2 = \frac{V_s}{R_2} \tag{4.35b}$$

If we replace the two resistors by a single equivalent one of value R_t, the current taken by the new resistor, I_t, must be the same as that taken by the parallel combination. Thus,

$$I_t = \frac{V_s}{R_t} \tag{4.36}$$

To be equivalent, I_t must be the sum of the individual currents. So,

$$I_t = I_1 + I_2$$

or,

$$\frac{V_s}{R_t} = \frac{V_s}{R_1} + \frac{V_s}{R_2}$$

Thus,

$$\frac{1}{R_t} = \frac{1}{R_1} + \frac{1}{R_2} \tag{4.37}$$

So, we can decrease resistance by adding another resistor in parallel with the original.

Let us now consider a series combination of resistors as shown in Fig. 4.7(b). As before, we will connect this combination to a d.c. source, and replace the resistors by a single equivalent one. Now, each of the individual resistors will have a potential across them, but pass the same current. Thus

$$V_1 = I R_1 \tag{4.38a}$$

and,

$$V_2 = I R_2 \tag{4.38b}$$

These individual potentials will add to give the supply voltage. So,

$$V_s = V_1 + V_2 \tag{4.39}$$

However, our equivalent resistor will have a potential across it given by

$$V_s = I R_t \tag{4.40}$$

So, by combining Equations (4.40), (4.39) and (4.38a,b) we get

$$I R_t = I R_1 + I R_2$$

resulting in

$$R_t = R_1 + R_2 \tag{4.41}$$

Thus we can increase resistance by adding another resistor in series with the original.

Example 4.6

A 100 kΩ resistor is connected in a circuit. What is the effect of placing a 1 kΩ resistor in parallel with it? What happens if the 1 kΩ is connected in series with the original?

Solution

We have a 100 kΩ resistor, and a 1 kΩ connected in parallel. So, the total resistance is

$$\frac{1}{R_t} = \frac{1}{R_1} + \frac{1}{R_2}$$

$$= \frac{1}{100} + \frac{1}{1}$$

$$= 1.01$$

i.e.

$R_t = 0.99$ kΩ

So the 1 kΩ resistor dominates the 100 kΩ resistor when they are in parallel.
If we now connect the 1 kΩ resistor in series, we get a new resistance of

$R_t = 100 + 1$
$= 101$ kΩ

Thus the resistance has barely altered at all.

4.5 SOME APPLICATIONS

In general, electroconduction is the most familiar of all the field systems we have considered. Some typical applications include resistors, conducting wire, fuse wire and heating elements in electric fires, cookers and immersion heaters. As these examples are so familiar, we will not consider them. Instead we will examine the formation of a resistor in semiconductor material.

When designing analogue circuits, resistors are often used. If the circuits are fabricated on printed circuit board, the designer can choose between wire-ended or surface-mount resistors. When producing an analogue integrated circuit, designers have no choice but to fabricate the resistors out of semiconductor material.

Figure 4.8 shows the basic form of an integrated circuit (i.c.) resistor. It is

140 Electroconductive fields

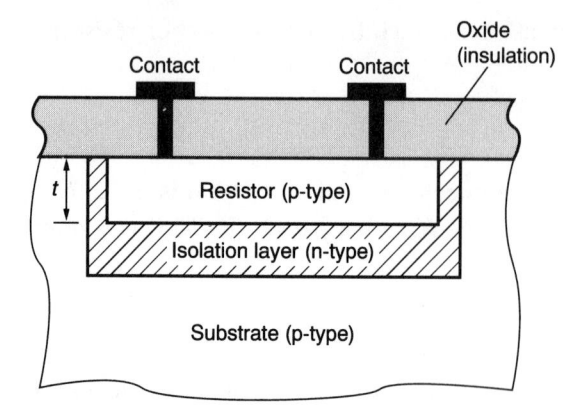

Fig. 4.8 A basic integrated circuit resistor

formed by depositing electrodes on to the silicon, and doping the material to the required conductivity. (Doping involves diffusing impurities into the silicon to alter the material conductivity.)

When working with i.c. resistors, the surface resistance of the material is used. If we consider a square of material, with electrodes on opposite sides, the resistance will be given by

$$R = \frac{l}{\sigma \times \text{area}}$$

where l is the length of one side of the square,
σ is the material conductivity, and
area is the cross-sectional area of the resistor.

Thus,

$$R = \frac{l}{\sigma \times l \times t}$$

$$= \frac{1}{\sigma \times t} \tag{4.42}$$

where t is the thickness of the resistor. So, the resistance of a square of material is independent of the surface area. This is a very important result because it means that i.c. designers can reduce the size of their resistors without altering the resistance.

The precise resistance per square depends on the impurity diffusion and thickness, but usually ranges from 2 Ω/square to 200 Ω/square. So, if we have a resistor with dimensions 1 by 10, we have 10 squares giving a resistance of 20 Ω to 2 kΩ.

4.6 SUMMARY

This chapter has been concerned with probably our most familiar field systems – electroconductive fields. We began by examining what we mean by current flow. We then went on to discuss the use of potential and electric field strength. This led us on to a discussion of current density, and introduced us to the fundamental relationship

$$J = \sigma E \tag{4.43}$$

We were then introduced to Ohm's law,

$$\frac{V}{I} = R \tag{4.44}$$

and the idea of resistance,

$$R = \frac{l}{\sigma A} \tag{4.45}$$

The resistance of a parallel plate capacitor, and various transmission lines, was then examined. The relevant resistances are summarized here:

capacitor $\quad\quad R = \dfrac{d}{\sigma A} \, \Omega \quad \text{(shunt)} \tag{4.46}$

coaxial cable $\quad R = \dfrac{l}{\sigma A} \, \Omega \quad \text{(series)} \tag{4.47}$

and $\quad\quad\quad\quad R = \dfrac{\ln(b/a)}{2\pi\sigma \times \text{length}} \, \Omega \quad \text{(shunt)} \tag{4.48}$

twin feeder $\quad\quad R = \dfrac{l}{\sigma A} \, \Omega \quad \text{(series)} \tag{4.49}$

microstrip $\quad\quad R = \dfrac{l}{\sigma A} \, \Omega \quad \text{(series)} \tag{4.50}$

We also came across a very important relationship: the capacitance and resistance are related by

$$RC = \frac{\epsilon_0 \epsilon_r}{\sigma} \tag{4.51}$$

Thus we can find the resistance if we know the capacitance.

One of the surprising aspects of this work is that a capacitor can also have resistance. This is due to imperfect dielectrics. (In reality all dielectrics will allow some current to flow.) The loss tangent characterizes this loss as the ratio of conduction to displacement current, i.e.

$$\tan \delta = \frac{I_{con}}{I_{disp}} \tag{4.52}$$

or,

$$\tan \delta = \frac{\sigma}{\omega \epsilon} \tag{4.53}$$

So, the lower the value of loss tangent the better the dielectric. An alternative viewpoint is that the higher the proportion of displacement current, the better the dielectric.

Next we examined parallel and series combinations of resistors. We saw that resistance can be decreased by adding another resistor in parallel with the original, and increased by adding series resistance.

We concluded this chapter with an examination of resistance in integrated circuits. We found that the resistance depends on the number of 'squares' between the resistor contacts. Thus quite large resistors can be physically very small.

5 Comparison of field equations

We have now completed our study of electrostatic, electromagnetic, and electroconductive fields. In this chapter we will compare and contrast the three field systems we have met. This should prove instructive, and reveal some interesting facts. The remaining two chapters discuss the physical effects of dielectrics on capacitors, and iron cores on transformers.

We will begin our comparison by examining the creation of a force field in the three systems we have considered.

5.1 FORCE FIELDS

In all the field systems we have considered, we started from the basic premise that like charges or poles repel, and unlike charges or poles attract. This led to the observation that the force decreases as the square of the distance between the charges or poles, i.e.

electrostatics $$F = \frac{q_1 q_2}{4\pi \epsilon r^2}$$

magnetostatics $$F = \frac{p_1 p_2}{4\pi \mu r^2}$$

electroconduction $$F = \frac{q_1 q_2}{4\pi \epsilon r^2}$$

As we can see, there is a great deal of similarity in the general forms of these equations – they all describe a force field of some sort. We should note that these equations assume point sources and, as we have seen with magnetism, this can be open to question. In spite of this fact, this does help us to visualize the field systems.

Once we accept that we are studying a force field, it is a simple step forward to define field strength as the force on a unit charge or pole. Thus, electrostatics $\quad E = \dfrac{q_1}{4\pi\epsilon r^2}$

magnetostatics $\quad H = \dfrac{p_1}{4\pi\mu r^2}$

electroconduction $\quad E = \dfrac{q_1}{4\pi\epsilon r^2}$

Let us take a moment to study these equations in detail. We should be fairly happy with the force field in an electrostatic system – we have all rubbed a balloon and felt the hairs on the back of our hands stand up. Anyone who has played with permanent magnets has seen the effect of the force field in a magnetic system. However, the idea of a force field in electroconductive systems is rather difficult to accept. This is because we are generally taught, at an early stage, a model of current flow that has the potential of a source as the force that drives current around a circuit. This idea is often reinforced by the term electromotive force for the potential of a source. (A water analogy is also often used. Although this analogy can be useful, its pull is very seductive, and we must be very careful.) As this model is very simple, it is generally very hard to accept that it is wrong. Potential does not force current around a circuit: it is the electric field set up in a conductor that causes current to flow.

So, each field system has a force field that repels or attracts charges or poles. This raises the question of what radiates from the charge or pole generating the force field. This is where we introduce the ideas of flux and flux density.

5.2 FLUX, FLUX DENSITY AND FIELD STRENGTH

In Chapter 2 we came across flux as the flow of material from one place to another. So, if a charge radiates flux, it should run out at some time! (After all, if material transfers from one place to another, the source will eventually run out.) This is where we must exercise extreme caution – flux is only a construct to help us visualize the field system. In electrostatic and magnetostatic field systems, no flux physically flows. In spite of this statement, in an electroconductive field system, the flow of flux does signify the transfer of charge, and that is a physical phenomenon. (Well, there's always an exception to the rule!) So, a source radiates flux, and a sink attracts it. A little thought shows that the flux depends on the force field – a large force field implies a large flux. Let us now examine this in greater detail.

In all the field systems we considered, we met a flux density of some sort: electric flux density in electrostatics; magnetic flux density in electromagnetism; and current density in electroconduction. In each of these systems, material constants related the flux densities to the respective field strengths, i.e.

electrostatics $\quad D = \epsilon E$

where $\quad D = \dfrac{\psi}{\text{area}} \quad$ and $\quad E = \dfrac{V}{l}$

electromagnetism $\quad B = \mu H$

where $\quad B = \dfrac{\phi}{\text{area}} \quad$ and $\quad H = \dfrac{NI}{l} = \dfrac{V_m}{l}$

electroconduction $\quad J = \sigma E$

where $\quad J = \dfrac{I}{\text{area}} \quad$ and $\quad E = \dfrac{V}{l}$

This comparison implies a fundamental relationship:

flux density = material constant × field strength

This is a quite remarkable result. Here we are with three different field systems, all of which have the same simple fundamental relationship. We should also note that these equations are spatial in form – they use the flux density, not the actual flux.

We can also observe that the field strength (or potential gradient) is given by

field strength = $\dfrac{\text{potential}}{\text{path length}}$

where potential is the work done in moving a unit test charge or pole around the path length. So, the spatial equations are all similar in form, but what about the 'point forms' of these equations?

5.3 POTENTIAL AND RESISTANCE TO FLUX

In the last section we compared the three fundamental laws governing electrostatic, electromagnetic and electroconductive fields. We can produce the point form of these laws by substituting for the flux densities and field strengths to give, after some rearranging,

electrostatics $\quad V = \dfrac{\psi}{C}$

where $\quad \dfrac{1}{C} = \dfrac{l}{\epsilon \times \text{area}}$

146 Comparison of field equations

electromagnetism $\quad V_m = \phi S$

where $\quad S = \dfrac{l}{\mu \times \text{area}}$

electroconduction $\quad V = IR$

where $\quad R = \dfrac{l}{\sigma \times \text{area}}$

Again there is a large degree of similarity among the three field systems. In particular, we can say:

potential = flux × resistance to flux

or,

work done = flux × resistance to flux

where the potential is the work done in moving a unit charge or pole around a path in a force field.

Also, the resistance to the flow of flux in all the systems is governed by

$$\text{resistance to flux} = \dfrac{\text{length of circuit}}{\text{material constant} \times \text{area}}$$

5.4 ENERGY STORAGE

Of the three field systems we have considered, only the electrostatic and electromagnetic fields can store energy – the electroconductive system dissipates energy as current flows. From a 'circuits' point of view, the stored energy is given by

electrostatics $\quad$ energy $= \tfrac{1}{2}CV^2$ joule

electromagnetism $\quad$ energy $= \tfrac{1}{2}LI^2$ joule

Now, capacitors rely on the voltage across them to generate an electric field, whereas inductors use the current through them to produce a magnetic field. With this in mind, we can see a great deal of similarity between the two circuit elements. This similarity is even greater if we compare the energy stored in the fields:

electrostatics $\quad$ energy $= \tfrac{1}{2}DE$ J m^{-3}

electromagnetism $\quad$ energy $= \tfrac{1}{2}BH$ J m^{-3}

So, in general we can say:

energy density $= \tfrac{1}{2} \times$ flux density $\times$ field strength

Let us just take a moment to compare the efficiency of both methods of storing energy. Let us consider a capacitor made of two circular plates, of area 5 cm², separated by 3 μm of glass with ϵ_r of 10. If the potential between the plates is 100 V, the stored energy is

$$\text{energy} = \frac{1}{2} \frac{14.8 \times 10^{-9}}{5 \times 10^{-4}} \frac{\times 100^2}{\times 3 \times 10^{-6}}$$

$$= 49 \text{ kJ m}^{-3}$$

In order to attain the same energy density, the H field in an inductor with a mumetal core, $\mu_r = 1 \times 10^5$, would need to be

$$H^2 = \frac{2 \times 49 \times 10^4}{4\pi \times 10^{-7} \times 1 \times 10^5}$$

and so,

$H = 883$ At m^{-1}

If we take a coil length of 10 cm, and pass 100 mA through the coil, this implies 883 turns of wire. If we assume that the capacitor and inductor are air cored, the inductor would need 88.3×10^3 turns for the same length! Obviously capacitors are better at storing energy.

5.5 FORCE

Both the electrostatic and magnetic field systems can be used to apply force to charged or magnetic surfaces. We were able to calculate the force by using the energy stored per unit volume, and moving one of the surfaces a very small amount. This gave us the following results:

electrostatics   force $= \frac{1}{2} \psi E$ newton

electromagnetism   force $= \frac{1}{2} \phi H$ newton

Thus we can say:

force $= \frac{1}{2} \times$ flux $\times$ field strength

As with the previous section, let us compare the relative efficiencies of electromagnetic and electrostatic lifting equipment. Let us initially consider a horseshoe shaped piece of iron, of circular cross-sectional area 10 cm², and overall length 60 cm. To convert this to an electromagnet, two coils of 750 turns each are wound on each limb, and connected in series. Let us try and find the current required to lift 200 kg masses, if an air gap of 0.1 mm is present at each pole.

Now, first we must find the force required to lift these weights. This lifting force must equal that due to gravity. So,

$F = ma$
$= 200 \times 9.81$
$= 1.962 \times 10^3$ newton

This force must be generated by the electromagnet and so,

$1.962 \times 10^3 = \frac{1}{2} \phi H \times 2$

$= \frac{1}{2} B \times \text{area} \times \dfrac{B}{\mu_0} \times 2$

(The factor of two appears in this equation because we effectively have two pole faces.) Thus,

$B^2 = \dfrac{1.962 \times 10^3 \times 4 \times \pi \times 10^{-7}}{10 \times 10^{-4}}$

$= 2.466$

or,

$B = 1.57 \text{ Wb m}^{-2}$

This is the flux density in the air gap as well as that in the former of the electromagnet. If we take a μ_r of 410, we get a magnetic field strength of 3×10^3 At m^{-1} in the former. If we ignore the reluctance of the air gap and the metal being lifted, we get

$H = \dfrac{NI}{l}$

and so,

$NI = 3 \times 10^3 \times 60 \times 10^{-2}$
$= 1.8 \times 10^3$ At

Thus the current through the two coils of 750 turns each is

$I = \dfrac{1.8 \times 10^3}{2 \times 750}$

$= 1.2$ ampere

This is a quite respectable current, and one that a voltage source can supply without too much trouble.

Let us now turn our attention to the use of electrostatics. To make a fair comparison, we will assume that we have two metal plates of diameter equal to the spacing between the poles of the electromagnet. The use of a steel rule, and a little imagination, means we can take a plate diameter of 25 cm. If we assume the plates are separated by 0.1 mm as before, and we wish to lift 200 kg again, we get

$$1.962 \times 10^3 = \tfrac{1}{2} \psi E$$
$$= \tfrac{1}{2} \epsilon_0 E \times \text{area} \times E$$

and so,

$$E^2 = 9 \times 10^{15}$$

or,

$$E = 9.5 \times 10^7 \text{ V m}^{-1}$$

This results in a potential between the plates of

$$V = 9.5 \times 10^7 \times 0.1 \times 10^{-3}$$
$$= 9.5 \text{ kV}$$

This is a rather large voltage to say the least! In reality, air breaks down at a field strength of 3×10^6 V m^{-1} and so we can't use electrostatics here.

So, to summarize, electrostatic fields are more efficient at storing energy than electromagnetic fields. However, electromagnetic fields are more efficient at lifting materials.

5.6 SUMMARY

We have seen a great deal of commonality among our three different field systems. The fundamental relationships we found are summarized here:

flux density = material constant × field strength

$$\text{field strength} = \frac{\text{potential}}{\text{path length}}$$

potential = flux × resistance to flux

$$\text{resistance to flux} = \frac{\text{length of circuit}}{\text{material constant} \times \text{area}}$$

energy density = $\tfrac{1}{2}$ × flux density × field strength

force = $\tfrac{1}{2}$ × flux × field strength

We have now finished our comparison of our three field systems. In the next chapter we will examine what happens to dielectrics in an electric field.

6 Dielectrics

In this chapter we will re-examine capacitors and, in particular, the effect of an electric field on dielectrics. As we saw in Chapter 2, dielectrics have a permittivity greater than air. Thus, when we use them as the insulating material in a capacitor, they increase the capacitance. If the potential across the plates is constant, the stored charge will increase and so it is generally desirable to design a capacitor with a dielectric.

The next section introduces the dipole moment that occurs when an electric field distorts an atom, or molecule.

6.1 ELECTRIC DIPOLES AND DIPOLE MOMENTS

If we have an atom that is well away from any external electric field, the orbit of the electrons will describe a sphere with the nucleus at the centre (see Fig. 6.1(a)). If we place this atom in an electric field, the field will distort the atom as Fig. 6.1(b) shows. In effect, the electron cloud moves away from the field, whereas the positively charged nucleus moves in the direction of the field. The atom is said to be polarized by the electric field. Although the polarized atom is still electrically neutral, on the microscopic scale the atom has an electric field. This becomes clearer if we replace the electron cloud by a point source at a distance d from the positive nucleus. This arrangement, shown in Fig. 6.1(c), is known as an electric dipole.

So, when we subject an electrically neutral atom to an external electric field, we can represent the atom by an electric dipole. We tend to characterize electric dipoles by the dipole moment, p, given by

$$p = q \times d \text{ C m} \tag{6.1}$$

where q is the charge on the nucleus or electron cloud, and d is the effective separation of the two charges. The dipole moment is a vector quantity with direction towards the positive charge. The reason for this choice of direction will become clear in the next section.

In most materials, the separation between the charges is directly dependent on the magnitude of the external E field. However, if we continually increase the E field, there comes a point where we cannot polarize the atom any more.

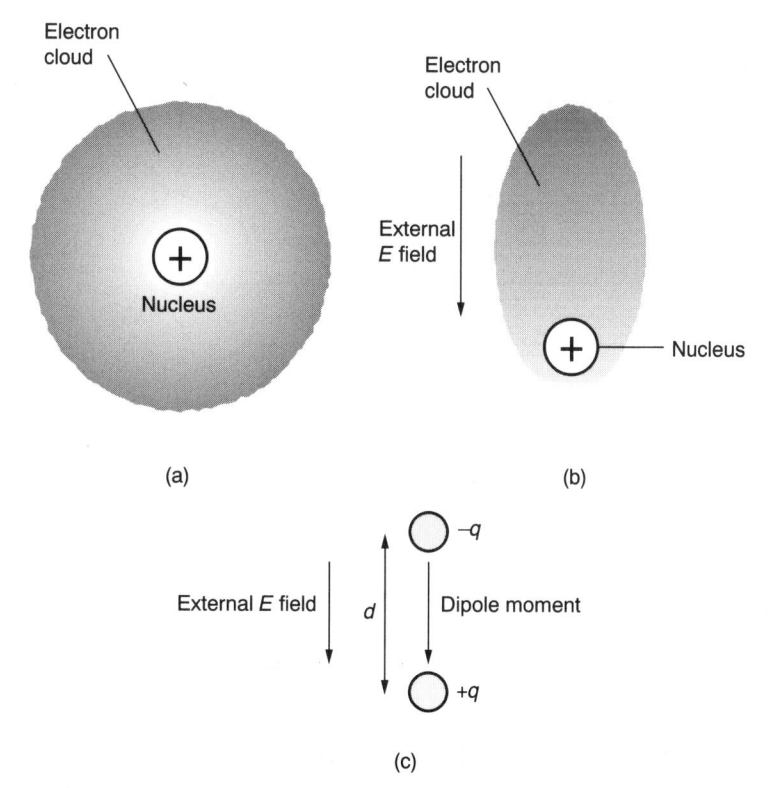

Fig. 6.1 (a) Structure of an atom with no external E field; (b) polarization of atom due to external E field; and (c) equivalent electric dipole

The material is then said to be saturated. If we increase the E field beyond this point, the force on the dipole becomes so great that it breaks apart and becomes ionized. If the dielectric is the insulating material in a capacitor, this causes a small, but very dramatic, explosion! So, it is important to know the field at which the dielectric breaks down. This is the subject of Section 6.4.

Some materials exhibit dipole moments on a molecular level, or macroscopic scale. Figure 6.2 shows a charge embedded in some insulating material. As this charge is negative, it tends to polarize the surrounding molecules producing molecular dipole moments. The net effect is that these molecular dipoles are able to exert a force on each other, and on any external charges.

Now, what happens if we introduce a test charge into this region? Logically each dipole moment will exert a force on the test charge. However, the net force will be less than if the material was not there. This is because the force acting on the test charge is due to the vector sum of the dipole moments. As the distribution of these dipoles is random throughout the material, the net result is a smaller force than if the material was not present.

An alternative explanation is to consider the direction of the electric field produced by each dipole. As Fig. 6.2 shows, these fields act in the opposite

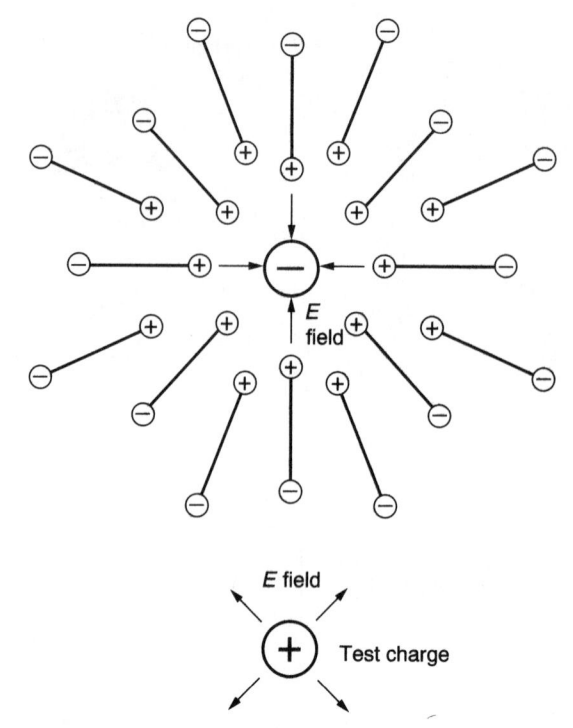

Fig. 6.2 Effect of charge on molecular electric dipoles

direction to the point charge field, so reducing the field from the point charge. We can check the validity of this model by noting

$$E = \frac{D}{\epsilon} \tag{6.2}$$

So, if the charge is in air, the force field is

$$E = \frac{D}{\epsilon_0} \tag{6.3}$$

However, if the charge is in material with a relative permittivity of ϵ_r, the force field is

$$E = \frac{D}{\epsilon_0 \epsilon_r} \tag{6.4}$$

which is lower than for air. This implies that it is the polarization of the material that causes a permittivity greater than air. This is the subject of the next section.

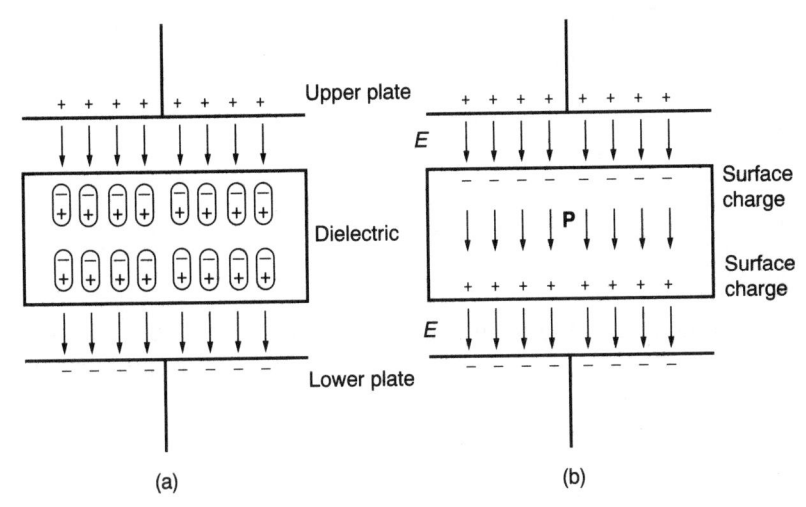

Fig. 6.3 (a) Dielectric-filled capacitor; and (b) macroscopic view of dielectric-filled capacitor

6.2 POLARIZATION AND RELATIVE PERMITTIVITY

We have just seen that the material in which a charge is placed reduces the electric field strength produced by a point charge. What we have not yet examined is the effect of a dielectric on a capacitor.

Figure 6.3(a) shows a capacitor with a dielectric between its plates. When the upper plate has a positive charge on it, flux radiates down across the air gap and into the dielectric. The E field in the capacitor will polarize the dielectric, so producing a negative charge on the upper surface of the dielectric. A similar situation occurs on the lower plate, and so we can take the macroscopic view as Fig. 6.3(b) shows. So, the external field induces polarization in the dielectric with the dipole moments pointing towards the lower plate. This implies a polarization vector which we can define as follows.

Let us consider a small volume of dielectric, δV, that is large enough to enclose a large number of atoms. If the number of atoms per unit volume is N, the number of atoms in this volume is $N\delta V$. Now, let us assume that all of these atoms are polarized in the same direction – acting downwards. Each atom has a dipole moment of

$$p = q \times d \text{ C m} \tag{6.5}$$

and so the total polarization of the volume δV is

$$N\delta V \times qd \text{ C m} \tag{6.6}$$

Thus the polarization per unit volume, P, is

$$P = \frac{\text{polarization of } \delta V}{\delta V}$$

$$= \frac{N\delta V \times qd}{\delta V}$$

which gives

$$P = Nqd \text{ C m}^{-2} \tag{6.7}$$

So, we have a new vector, the polarization, with units of coulomb per m² acting in the same direction as the **E** field. This means we can modify our equation linking **D** and **E** to give

$$\boldsymbol{D} = \epsilon_0 \boldsymbol{E} + \boldsymbol{P} \tag{6.8}$$

Materials in which the dipole moment is directly proportional to the *E* field are known as linear materials, and most dielectrics we come across are linear in form. Thus we can write

$$\boldsymbol{P} = \chi \epsilon_0 \boldsymbol{E} \tag{6.9}$$

where χ is a constant of proportionality called the electric susceptibility of the material. We can now write Equation (6.8) as

$$\boldsymbol{D} = \epsilon_0 \boldsymbol{E} + \chi \epsilon_0 \boldsymbol{E}$$
$$= (1 + \chi)\epsilon_0 \boldsymbol{E}$$

Thus the relative permittivity of the material is

$$\epsilon_r = 1 + \chi \tag{6.10}$$

i.e. ϵ_r depends on the electric susceptibility of the material, which is itself a measure of the polarizability of the dielectric.

Example 6.1

Two metal plates of area 10 cm² are placed 5 mm apart. A dielectric, with $\epsilon_r = 2.5$ and thickness 3 mm, is placed mid-way between the plates. A potential of 200 V is applied across the plates. Determine:

(1) the capacitance of the arrangement
(2) the surface charge densities
(3) the flux density at all points
(4) the electric field strength at all points
(5) the bound charge densities.

Solution

Figure 6.4(a) shows the situation we seek to analyse. As this figure shows, there is an air gap both above and below the dielectric. Now, by considering the

6.2 Polarization and relative permittivity

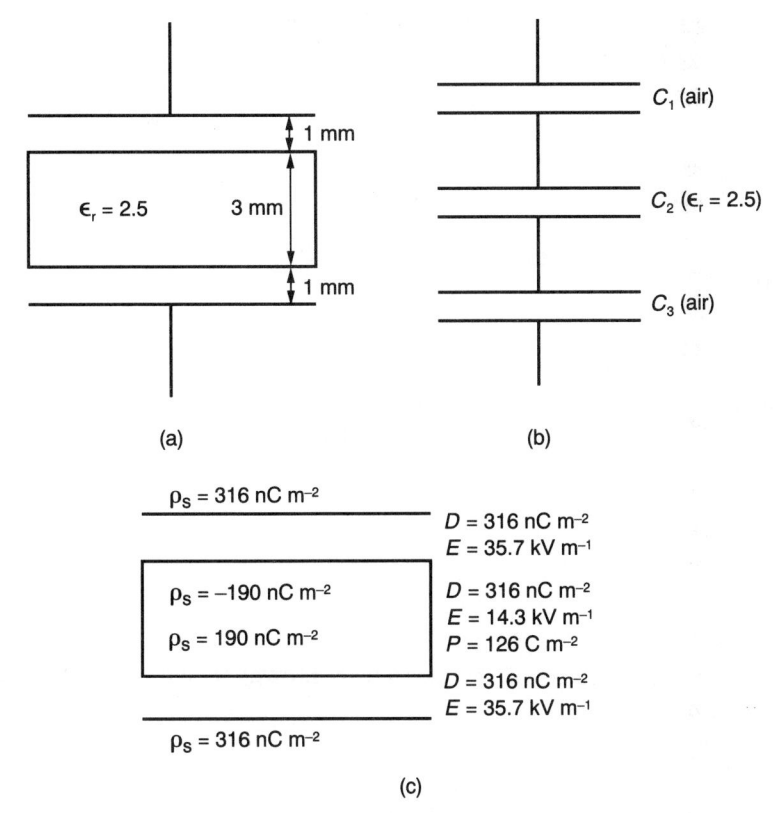

Fig. 6.4 (a) Schematic of composite dielectric capacitor; (b) equivalent circuit of composite dielectric capacitor; and (c) distribution of flux density and surface charges

charge distribution throughout the capacitor we can see that we can replace the composite capacitor by three capacitors in series (Fig. 6.4(b)).

We can find the total capacitance from

$$\frac{1}{C_t} = \frac{1}{C_1} + \frac{1}{C_2} + \frac{1}{C_3}$$

$$= \frac{1 \times 10^{-3}}{\epsilon_0 \times 10 \times 10^{-4}} + \frac{2 \times 10^{-3}}{\epsilon_0 \times 2.5 \times 10 \times 10^{-4}} + \frac{1 \times 10^{-3}}{\epsilon_0 \times 10 \times 10^{-4}}$$

$$= 3.16 \times 10^{11}$$

and so,

$C_t = 3.16$ pF

The voltage across the capacitor is 100 V and so the stored charge is

$Q_t = 3.16 \times 10^{-12} \times 100$
  $= 316$ pC

Thus the top plate will have a charge of 316 pC on it giving a surface charge density of

$$\rho_s = \frac{316 \times 10^{-12}}{10 \times 10^{-4}}$$
$$= 316 \text{ nC m}^{-2}$$

Each unit charge will radiate a unit of flux, and so the flux density in the air gap will be

$$D = 316 \text{ nC m}^{-2}$$

This flux density will remain constant throughout the capacitor.

With this flux density, the electric field strength in each air gap is

$$E_a = \frac{316 \times 10^{-9}}{8.854 \times 10^{-12}}$$
$$= 35.7 \text{ kV m}^{-1}$$

Let us now turn our attention to the dielectric. We have a flux density of 316 nC m^{-2} at the surface of the capacitor. However, we have just seen that (Equation (6.8))

$$\boldsymbol{D} = \epsilon_0 \boldsymbol{E} + \boldsymbol{P}$$

or,

$$\boldsymbol{D} = \epsilon_0 \epsilon_r \boldsymbol{E}$$

So, the $\boldsymbol{E}$ field in the dielectric is

$$E = \frac{D}{\epsilon_0 \epsilon_r}$$
$$= \frac{316 \times 10^{-9}}{8.854 \times 10^{-12} \times 2.5}$$
$$= 14.3 \text{ kV m}^{-1}$$

Now, the difference between the air and dielectric field strengths is due to the bound charges on the surface of the dielectric. So, as $\boldsymbol{D} = \epsilon_0 \boldsymbol{E}$, the surface charge density (or flux density) is

$$\rho_d = \epsilon_0 \times (14.3 - 35.7) \times 10^3$$
$$= -8.854 \times 10^{-12} \times 21.4 \times 10^3$$
$$= -190 \text{ nC m}^{-2}$$

Figure 6.3(c) shows the distribution of flux density and surface charges throughout the capacitor.

At first sight, the work we have done in this section may not appear to be of much practical benefit. After all, we are quite happy to work with the relative permittivity of a material, so why should we concern ourselves with polarization? The answer to this question should become clear in the next section when we examine what happens to an electric field as it crosses a dielectric boundary.

6.3 BOUNDARY RELATIONSHIPS

In the example at the end of the last section we encountered a composite dielectric capacitor. This meant that the electric field in the capacitor had to cross a boundary between two dielectrics. The question of what happens to the *E* field is very important – light and radio waves are electromagnetic in form, and it would be nice to know what happens to these signals as they pass through a dielectric.

Figure 6.5(a) shows an electric field crossing the boundary between two dielectrics. To analyse this situation we will split the incident and transmitted field into their tangential and perpendicular components.

Figure 6.5(b) shows the tangential component of the *E* field either side of the boundary. Let us consider a rectangular path ABCD that lies either side of the boundary. Now, the work done in moving a unit point charge from point A around the path and back again must equal zero. (We have moved the charge around a closed loop, and so have not gained or lost potential.) So,

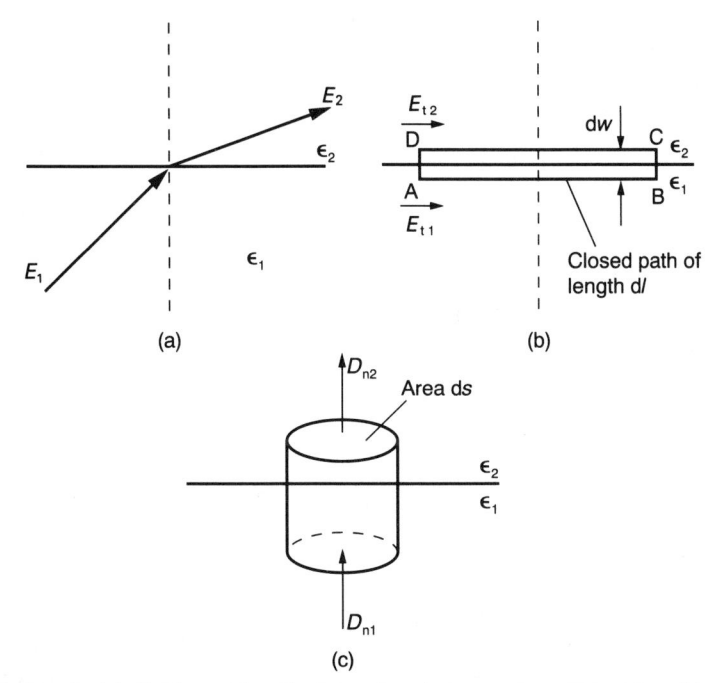

Fig. 6.5 (a) An electric field crossing the boundary between two dielectrics; (b) tangential *E* field at the boundary; and (c) normal component of *D* field at the boundary

$$\int_{ABCDA} E\, dl = E_{t1}\, dl + E_{n1}\, dw - E_{t2}\, dl - E_{n2}\, dw$$
$$= 0 \qquad (6.11)$$

If we make this path extremely thin, i.e. $dw \to 0$, we get

$$E_{t1}\, dl - E_{t2}\, dl = 0$$

and so,

$$E_{t1} = E_{t2} \qquad (6.12)$$

i.e. the tangential component of the E field is continuous across the boundary.

To find out what happens to the normal component of the fields, let us construct a pill-box with the top face in medium 2, and lower face in medium 1, as shown in Fig. 6.5(c). Now, electric flux entering the pill-box does so via the lower face, and flux leaving the pill-box does so via the upper face. If there are no surface charges, the flux entering must equal the flux leaving, i.e.

$$\psi_1 = \psi_2 \qquad (6.13)$$

or

$$D_{n1}\, ds = D_{n2}\, ds$$

and so,

$$D_{n1} = D_{n2} \qquad (6.14)$$

i.e. the normal component of the D field is continuous across the boundary if there are no bound charges.

As we have seen in the previous section, polarization causes bound charges to appear at the surface between the two dielectrics. These charges will radiate additional flux out of the pill-box, and so Equation (6.13) becomes

$$\psi_2 = \psi_1 + q_s \qquad (6.15)$$

where q_s is the surface charge. If we use the surface charge density, we can write

$$D_{n2}\, ds = D_{n1}\, ds + \rho_s\, ds$$

and so,

$$D_{n2} = D_{n1} + \rho_s \qquad (6.16)$$

i.e. the normal component of the D field is discontinuous across the boundary if there are bound charges. The amount of discontinuity is equal to the surface charge density.

Before we consider an example, let us digress slightly and examine what happens at the boundary between a conductor and air. Let us introduce some charges into the conductor. These charges will repel each other until they end up in equilibrium on the surface of the conductor. Thus we will have a surface charge density similar to that which we met in the last example. The fact that the charges are in equilibrium on the surface of the conductor means that the tangential component of the E field must be zero inside the wire.

As regards the surface charges, these will radiate flux in a direction normal to the surface. (If there were a tangential component to the flux, this would generate a tangential E field and that is not allowed!) So, we can say

$$E_t = 0 \qquad (6.17)$$

and,

$$E_n = \frac{\rho_s}{\epsilon_0} \qquad (6.18)$$

Exactly the same result applies if the conductor is placed in an E field. This is of some practical benefit – if the conductor is an earthed plate, the surface charge will dissipate to ground and so the E field will not pass through. This is the principle behind the earthed shield in coaxial cable.

Example 6.2

An E field makes an angle of 30° to the horizontal at the boundary between air and a dielectric. If the dielectric has $\epsilon_r = 6$, determine the angle of refraction. Assume that there are no bound charges.

Solution

The tangential component of the E field is continuous across the boundary. So, the tangential E field in the dielectric is

$$E_t = E \cos 30°$$
$$= \frac{2}{\sqrt{3}} E$$

As there are no bound charges at the surface, the normal component of the D field is continuous across the boundary. Thus,

$$D_n = D \sin 30°$$

or,

$$D_n = \epsilon_0 E \sin 30°$$

Now, the dielectric has $\epsilon_r = 6$ and so

$$6 \times \epsilon_0 \times E_n = \epsilon_0 E \sin 30°$$

or,

$$E_n = \frac{E}{6}\frac{1}{\sqrt{3}}$$

We can find the angle of refraction from

$$\tan \alpha = \frac{E_n}{E_t}$$

$$= \frac{1}{6\sqrt{3}} \times \frac{\sqrt{3}}{2}$$

$$= \frac{1}{12}$$

Thus the refracted E field makes an angle of 4.8° to the horizontal. We should expect this because the normal component of the E field has been significantly reduced, whereas the tangential component has hardly altered.

6.4 DIELECTRIC STRENGTH AND MATERIALS

Although this is the last section in this chapter, it is probably the most important from an engineering point of view. As we have already seen, an electric field tends to polarize the atoms in a dielectric. If the field is great enough, the dielectric could break down by direct ionization of the atoms. This causes the leakage current to rise dramatically, resulting in a very rapid rise in temperature and an explosion! The dielectric strength of an insulator is the maximum field strength that the material can stand before ionization occurs. Table 6.1 lists the relative permittivities and dielectric strengths of some common insulating materials.

Table 6.1 Relative permittivity and dielectric strength for a variety of dielectric materials

Material	ϵ_r	Dielectric strength (V m^{-1})
Air (atmospheric pressure)	1.0	3×10^6
Mineral oil	2.3	15×10^6
Rubber	2.3–4.0	25×10^6
Paper	2.0–4.0	15×10^6
Polystyrene	2.6	20×10^6
Glass	4.0–10	30×10^6
Mica	6.0	200×10^6

A glance at this table shows a wide spread in the maximum field strength that insulators can withstand. However, they are all of the order of several million volts per metre. It may be thought that such a high E field will not occur in

6.4 Dielectric strength and materials

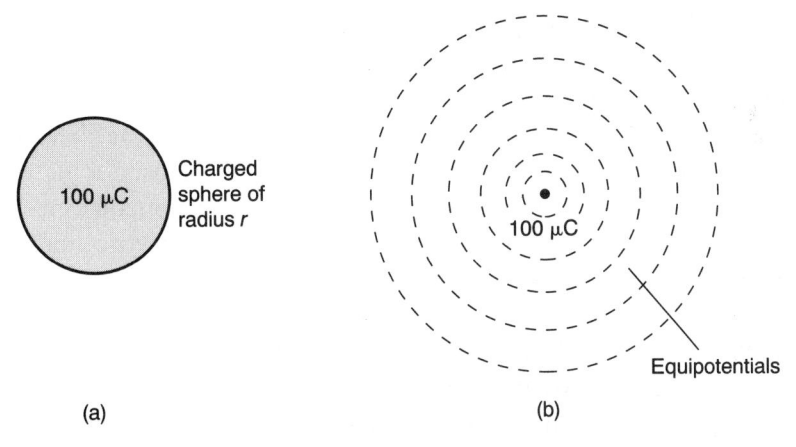

Fig. 6.6 (a) A charged sphere; and (b) the equipotential plot

practice. However, as we have already seen, the field inside a capacitor does approach the maximum listed in the table.

So, we must use care when designing a capacitor not to produce too large a field strength. What about the shape of the conductors? Is there one particular shape that results in the lowest field strength? To answer these questions let us consider a charged sphere, and calculate the E field at the surface.

Figure 6.6 shows a sphere that has a total charge of 100 μC on the surface. We can use Gauss' law to replace this sphere by a point charge of magnitude 100 μC at the centre of the sphere. Thus the E field at the surface is simply

$$E_r = \frac{100 \times 10^{-6}}{4\pi\epsilon_0 r^2}$$

$$= \frac{9 \times 10^5}{r^2} \tag{6.19}$$

where r is the radius of the sphere.

So, if we halve the radius of the sphere, the surface E field increases by a factor of 4. Clearly there will come a point where the E field will exceed the dielectric strength of air. Taking a maximum E field of 3×10^6 V m^{-1} gives a minimum radius of 55 cm. If we make the sphere any smaller, we could ionize the surrounding air, resulting in corona discharge of the sphere. (The fact that small conductors generate high field strengths is put to good use in lightning conductors. These have a pointed end that produces a large electric field. When the field is great enough, a corona discharge is created (the lightning strike), and charge flows along the lightning conductor.)

To guard against ionization in a capacitor, we must ensure that the capacitor plates are as uniform as possible – any points on the plates will generate a large electric field. So, the metal plates used in capacitors must be flat.

Two types of dielectric are currently used in capacitors: non-polar materials and polar materials. Non-polar materials are those listed in Table 6.1. Such dielectrics are used to give capacitors of value up to 0.1 µF.

Large-value capacitors tend to use polar materials, or electrolytes. These materials conduct electricity, and so why use them in capacitors? The answer lies in the construction of polarized capacitors. One of the capacitor plates has a very thin layer of aluminium oxide deposited on it. A conducting electrolyte separates this oxide from the other plate, and so the dielectric thickness is simply the thickness of the oxide. Thus the capacitance of this arrangement can be high.

Anyone who has used electrolytic capacitors will know that they have to be connected the right way round. This is because the electrolyte will conduct in one particular direction. Thus if we incorrectly connect an electrolytic capacitor, a very large current flows, resulting in a large bang and clouds of vaporized electrolyte!

7 Ferromagnetic materials and components

In this chapter we will re-examine transformers, and the effect of a magnetic field on ferrous materials. As we saw in Section 3.11, ferrous cores cause the inductance of a coil to increase. Thus iron-cored inductors can be physically smaller than their air-cored counterparts. One effect of an iron core is that it provides a low reluctance path to magnetic flux. This reduces any flux leakage, and leads to a more efficient transformer.

In common with dielectrics, a magnetic field produces a magnetic dipole moment in the ferrous material, and this is where we begin our studies.

7.1 MAGNETIC DIPOLES AND PERMANENT MAGNETS

In the last chapter, we developed a model of dielectric polarization in which an external field distorted the atoms into electric dipoles. In magnetic materials we can use a similar model in which the material has a random distribution of tiny permanent magnets, or magnetic dipoles (see Fig. 7.1(a)).

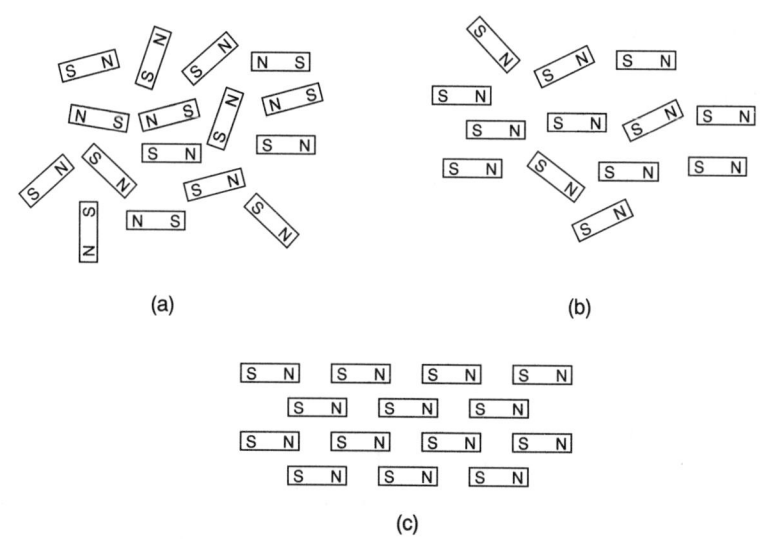

Fig. 7.1 (a) Randomly distributed magnetic dipoles; (b) distribution of magnetic dipoles due to external field; and (c) saturation of magnetic dipoles

Now, when we apply an external magnetic field to the material, some of the dipoles will line up in the direction of the field (Fig. 7.1(b)). If we continue to increase the external field, there comes a point when all the dipoles line up (Fig. 7.1(c)). This is known as saturation. If we now remove the external field, some of the magnetic dipoles will remain lined up in their new positions, so creating a permanent magnet.

Now, what happens if we introduce a unit pole into some magnetic material? This pole will tend to re-align some of the magnetic dipoles surrounding it. If we then introduce another pole at a certain distance from the original, this new pole will experience a force from the surrounding dipoles and not from the original monopole. Thus the material reduces the original force field. We can check the validity of this by noting that the magnetic force field in air is given by

$$H = \frac{B}{\mu_0} \tag{7.1}$$

If the pole is in a magnetic medium, the force field is

$$H = \frac{B}{\mu_0 \mu_r} \tag{7.2}$$

which is lower than for air. This implies that it is the polarization of the material that causes a permeability greater than air.

7.2 POLARIZATION AND THE *B/H* CURVE

When we discussed dielectrics in the previous chapter, we introduced the idea of dielectric polarization. As we have just seen, we have a similar situation in magnetic materials, and so we can introduce a magnetic polarization vector, **M**. Thus the flux density can be modified to

$$\boldsymbol{B} = \mu_0 \boldsymbol{H} + \boldsymbol{M} \tag{7.3}$$

Let us take a moment to examine this equation. If we have a coil wound on a ferrous former, the **H** field is directly proportional to the coil current. So, if we increase the current, the **B** field will increase. If the core is initially unmagnetized, a plot of *B/H* (Fig. 7.2) will follow curve OP. If we continue to increase the coil current, there will come a point at which the core saturates, point P. Thus further increases in current will not increase the flux density.

Let us now decrease the current. As **H** falls, the flux density will also decrease. Unfortunately **M** is not directly proportional to the field – this is due to some of the magnetic dipoles staying in their new positions. So, if we make **H** equal to zero, by removing the current, there will still be some residual magnetic flux. This is known as the remnance of the core, point Q in Fig. 7.2.

If we reverse the current through the coil, we need to produce an **H** field of

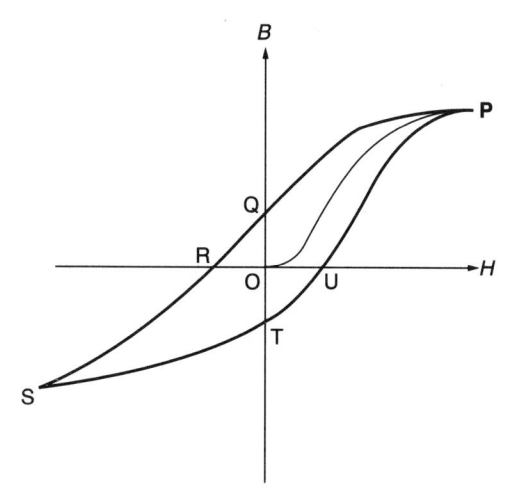

Fig. 7.2 B/H curve of a magnetic sample

OR to reduce the flux density to zero. This is known as the coercivity of the core. If we continue to increase the current we reach the reverse saturation point S. If we then start to reduce the current, we find that the B/H curve follows a new path, STUP. So we can never remove the magnetic effects completely. (We could heat the core up to the Curie temperature, at which point it does lose its remnance. However, this would cause the insulation surrounding the wire to melt, so destroying the coil!)

The fact that the B/H curve is not linear means that M is not directly proportional to B. Thus we cannot use the equivalent of electric susceptibility. Instead we have to accept that the relative permeability of the core is non-linear, and must be found from the B/H curve.

7.3 BOUNDARY RELATIONSHIPS

As stated in Section 6.3, light is an electromagnetic wave, i.e. it has an electric and a magnetic field. In that section we were interested in what happened at the boundary between two dielectrics. Here we will concern ourselves with the effect of a change in magnetic media on the *H* field.

Figure 7.3(a) shows a magnetic field crossing the boundary between two ferrous materials. To analyse this situation we will split the incident and transmitted fields into their tangential and perpendicular components.

Figure 7.3(b) shows the tangential component of the *H* field either side of the boundary. Let us consider a rectangular path ABCD that lies either side of the boundary. Now, Ampère's law states that the line integral of the *H* field around a closed path must equal the enclosed current. If we have no surface currents, we get

166 *Ferromagnetic materials and components*

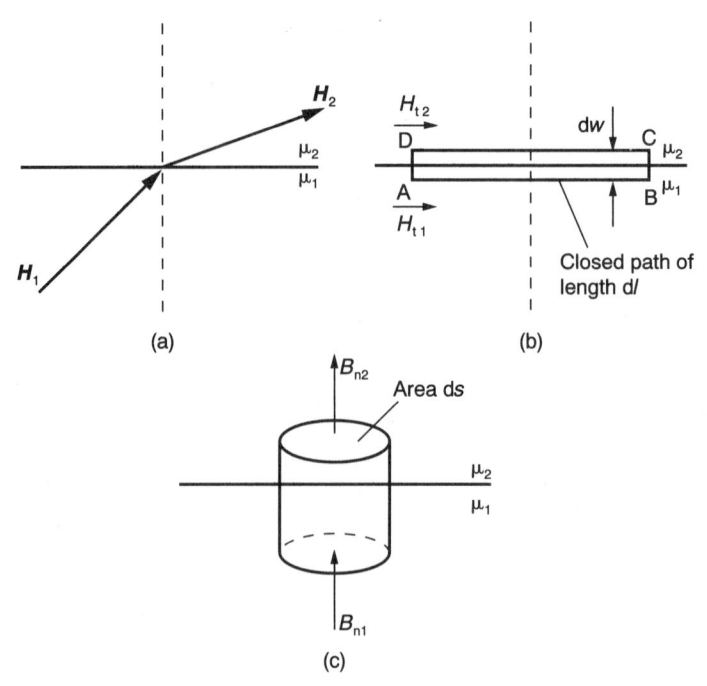

Fig. 7.3 (a) A magnetic field crossing the boundary between two different ferrous materials; (b) tangential **H** field at the boundary; and (c) normal component of **B** field at the boundary

$$\int_{ABCDA} H\, dl = H_{t1}\, dl + H_{n1}\, dw - H_{t2}\, dl - H_{n2}\, dw$$
$$= 0 \qquad (7.4)$$

If we make this path extremely thin, i.e. $dw \to 0$, we get

$$H_{t1}\, dl - H_{t2}\, dl = 0$$

and so,

$$H_{t1} = H_{t2} \qquad (7.5)$$

i.e. the tangential component of the *H* field is continuous across the boundary provided there are no surface currents.

It is interesting to note that if surface currents are set up in the new media, the transmitted **H** field will be severely attenuated and may even be zero for a good conductor. These surface currents are known as eddy currents. Coaxial cable uses this property to induce eddy current in the shielding. As the shielding is grounded, the eddy currents will be conducted to earth and lost.

To find out what happens to the normal component of the field, let us construct a pill-box with the top face in medium 2, and lower face in medium 1, as shown in Fig. 7.3(c). Now, magnetic flux entering the pill-box does so via the lower face, and flux leaving the pill-box does so via the upper face. As there can be no monopoles on the surface, the flux entering must equal the flux leaving, i.e.

$$\phi_1 = \phi_2 \tag{7.6}$$

or,

$$B_{n1} \, ds = B_{n2} \, ds$$

and so,

$$B_{n1} = B_{n2} \tag{7.7}$$

i.e. the normal component of the B field is continuous across the boundary.

7.4 IRON-CORED TRANSFORMERS

Iron-cored transformers use the fact that the flux density in ferrous material is greater than in air for the same H field. Such devices are found in mains-powered appliances and electrical substations. Figure 7.4 shows the general form of a transformer.

As can be seen from this figure, we have two coils wound on a former. Let us call the left-hand coil the primary, and the right-hand coil the secondary. Now, convention dictates that the primary is connected to the supply, and the secondary is connected to the load. As we have previously seen, an alternating voltage causes an alternating flux in the core. As the core has a permeability greater than that of air, the reluctance of the core is less than air – see Equation (3.38). So, most of the flux will be concentrated in the core. This alternating flux flows through the secondary coil where it generates an alternating voltage at the secondary terminals.

As we saw in Chapter 3, the primary voltage is given by (Equation (3.43))

$$V_1 = N_1 \frac{d\phi}{dt}$$

and so,

$$\frac{d\phi}{dt} = \frac{V_1}{N_1} \tag{7.8}$$

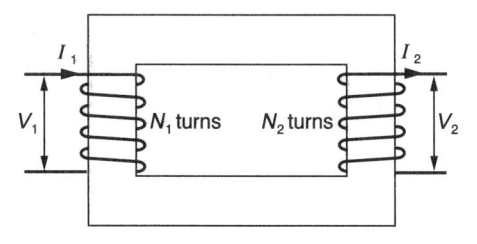

Fig. 7.4 Schematic of a typical transformer

This alternating flux generates a voltage at the secondary terminals of

$$V_2 = N_2 \frac{d\phi}{dt}$$

$$= N_2 \frac{V_1}{N_1}$$

i.e.

$$\frac{V_2}{V_1} = \frac{N_1}{N_2} \tag{7.9}$$

If N_1 is greater than N_2, the secondary voltage is less than the primary, i.e. we have a step-down transformer. If N_1 is less than N_2, the secondary voltage is greater than the primary and we have a step-up transformer. So, the turns ratio determines the secondary voltage, but what happens to the current?

Well, let us suppose that the transformer is ideal, i.e. there are no losses in the windings, and that the core has a linear B/H characteristic. Power taken by the secondary load has to be supplied by the primary and so

$$V_1 I_1 = V_2 I_2$$

By substituting for V_2 from Equation (7.9) we get

$$V_1 I_1 = V_1 \frac{N_2}{N_1} I_2$$

i.e.

$$\frac{I_2}{I_1} = \frac{N_1}{N_2} \tag{7.10}$$

Thus if we are considering a step-down transformer, the voltage will decrease by the turns ratio, but the current will increase by the same amount. The example at the end of this section shows this best.

Let us return to Equations (7.9) and (7.10). After some rearranging we get

$$V_2 = \frac{N_2}{N_1} V_1 \tag{7.11}$$

and

$$I_2 = \frac{N_2}{N_1} I_1 \tag{7.12}$$

If we divide Equation (7.11) by Equation (7.12) we get, after some rearranging,

$$R_1 = \left(\frac{N_1}{N_2}\right)^2 R_2 \tag{7.13}$$

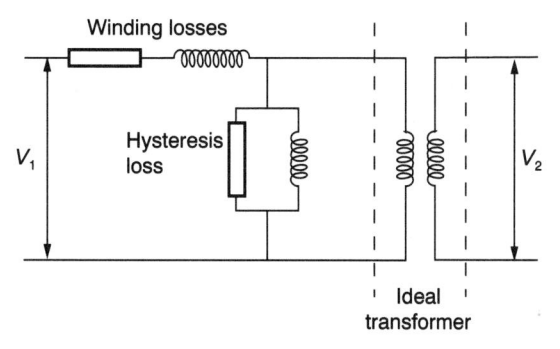

Fig. 7.5 Basic equivalent circuit of a transformer

So, if we have a step-down transformer, the secondary load appears to be greater from the primary side. This property means that we can use transformers to match source and load impedances for maximum power transfer.

Before we consider an example, let us take a moment to examine the losses that occur in an iron-cored transformer. As the transformer uses wire in the primary and secondary coils, there will be some resistive loss. In addition, the very fact that we are using coils means that there will be some primary and secondary inductance. As Equation (7.13) shows, we can refer the secondary losses to the primary as shown in Fig. 7.5. We can measure these losses, from the primary side, by performing a short-circuit test. We must not do these tests at the rated voltage because the high currents will destroy the transformer.

So, we can easily account for the winding losses. However, what about the shape of the *B/H* curve? Will this introduce additional loss? The answer to this question is yes. The *B/H* curve introduces a loss known as hysteresis loss. Let us refer to the *B/H* curve shown in Fig. 7.2. If the supply voltage is an alternating source, the current and **H** field will also alternate. This means that we go once round the *B/H* curve for every cycle of the supply current. We can find the energy stored in the magnetic field from

energy = $\int H \, dB$

which is simply the area of the *B/H* curve. So, we can find the hysteresis loss per cycle, and hence the power loss, if we know the area enclosed by the curve. If the area enclosed by the curve is small, the transformer is very efficient.

Example 7.1

An ideal transformer is wound on an iron former with 200 turns on the primary, and 30 turns on the secondary. Determine the secondary voltage if the primary is connected to a 240 V source. If a secondary load takes 5 A, determine the primary current and the equivalent load on the primary.

Solution

We have an ideal transformer with 200 turns on the primary, and 30 on the secondary. So, from Equation (7.9) we have

$$V_2 = \frac{N_2}{N_1} V_1$$

$$= \frac{30}{200} 240 = 36 \text{ volt}$$

If the secondary takes 5 A, the secondary power will be

$$P_2 = 36 \times 5 = 180 \text{ watt}$$

This power must come from the primary, and so

$$180 = V_1 \times I_1$$
$$= 240 \times I_1$$

Thus the primary current is 0.75 A.

Now, the secondary load takes 5 A from a 36 volt supply. So, the secondary load is

$$R_2 = \frac{36}{5} = 7.2 \, \Omega$$

The primary voltage is 240 V, and the primary current is 0.75 A. Thus the primary load is

$$R_1 = \frac{240}{0.75} = 320 \, \Omega$$

So, the 7.2 Ω secondary load appears to be a 320 Ω load when viewed from the primary.

7.5 ELECTRICAL MACHINERY

As we saw in Chapter 3, a force is exerted on a current-carrying wire placed in a magnetic field. So, what happens if we wind some turns on to a former that is free to rotate in a magnetic field?

Figure 7.6(a) shows a bobbin that is free to rotate on its axis. There are a number of turns on this former, which is placed between the poles of a permanent magnet. When we energize the coil, the coil current produces a magnetic field. This interacts with the field from the permanent magnet, so generating torque on the bobbin. This is known as the motor effect and it is used in moving-coil meters.

The force on one conductor is given by

$$F = BIl \sin \theta \tag{7.14}$$

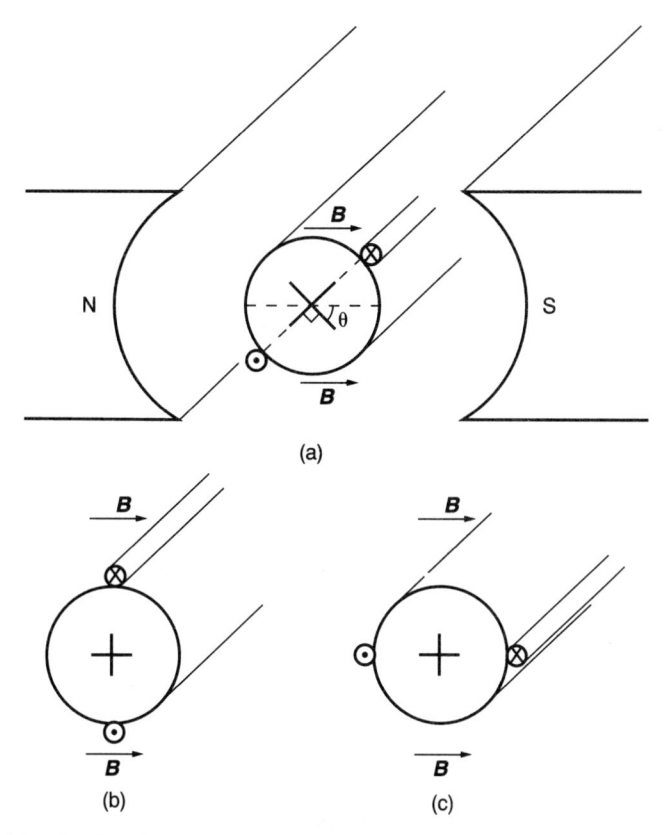

Fig. 7.6 (a) A basic electric motor; (b) position for minimum torque; and (c) position for maximum torque

where B is the flux density in the air gap, I is the current through the coil, l is the length of the bobbin, and θ is the angle between a line drawn normal to the axis of the coil and the field – see Fig. 7.6. (We came across a form of Equation (7.14) in Chapter 3 when we considered the Biot-Savart law. Yet again the angle θ is used as Equation (7.14) can be expressed using the vector crossproduct.)

Let us examine the factor $\sin \theta$ a bit more closely. If the coil is vertical (Fig. 7.6(b)), θ is 0° and, according to Equation (7.14), the rotational force is zero. If we draw the field due to the current in the wire, we can see that both conductors experience a force acting into the coil. Thus there is no rotational force on the wire.

If the coil is horizontal (Fig. 7.6(c)), θ is 90° and the rotational force should be a maximum. If we draw in the field generated by the current in the wire, we can see that the field lines on the left-hand conductor indicate an upward force, and those on the right-hand conductor show a downward force. So each conductor will experience maximum force.

We can turn this motor into a generator by driving the bobbin by some means. If we do this, the conductors will cut the flux from the permanent magnet, so inducing a voltage in the conductors. Specifically, the voltage induced in one conductor is

$$v(t) = B \times ld \times \omega \times \sin \omega t \tag{7.15}$$

where ld is the area of the coil, ω is the angular velocity of the bobbin, and the $\sin \omega t$ is there to account for the rotation of the coil. If we can extract this voltage by using slip rings, we can use it to supply current to an external load. Obviously, the prime mover that is rotating the bobbin has to supply this drain in power. This is similar to the transfer of power from a transformer primary to the secondary.

Example 7.2

A rectangular coil consists of $40\frac{1}{2}$ turns wound on a former that is 2 cm wide and 5 cm long. The coil is placed between the poles of a permanent magnet that generates a flux density of 10 mWb in the air gap. Determine the maximum torque on the coil if the current through it is 5 mA. If the coil is rotated at 100 revolutions per second, determine the voltage generated.

Solution

The former has $40\frac{1}{2}$ turns wound on it, i.e. there are 40 wires on the top of the coil, and 41 on the bottom. Now, the maximum torque occurs when the coil is horizontal. So, the force on one conductor is

$$F = BIl$$
$$= 10 \times 10^{-3} \times 5 \times 10^{-3} \times 5 \times 10^{-2}$$
$$= 2.5 \ \mu N$$

The torque on this single conductor is

$$\text{torque} = \text{force} \times \text{radius}$$
$$= 2.5 \times 10^{-6} \times 1 \times 10^{-2}$$
$$= 25 \times 10^{-9} \ N\,m$$

As we have 40 top conductors, and 41 bottom conductors, the total torque is

$$\text{torque} = (40 + 41) \times 25 \times 10^{-9}$$
$$= 2 \times 10^{-6} \ N\,m$$

If the motor is turned into a generator rotating at 100 revolutions per second, the voltage generated will be (Equation (7.15))

$$v = B \times ld \times \omega$$
$$= 10 \times 10^{-3} \times 2 \times 10^{-2} \times 5 \times 10^{-2} \times 2\pi \times 100$$
$$= 6.3 \ mV \text{ per conductor}$$

With 81 conductors, the terminal voltage will be

$v = 81 \times 6.3 \times 10^{-3}$
$= 0.5 \text{ V}$ at a frequency of 100 Hz.

So, the generator is not particularly good – 0.5 V will not do much. If we want to increase the voltage, we have to increase the dimensions of the generator and/or increase the flux density in the air gap.

7.6 THE MAGNETIC CIRCUIT

So far we have only considered simple magnetic structures. This enabled us to study the basics of magnetic fields. However, there are many complicated magnetic structures in use today. This is where we can make good use of an electrical analogy. The analysis of a magnetic circuit is best done by an example.

Figure 7.7(a) shows the cross-section through a relay that has energizing coils on the left-hand and centre limbs. The right-hand limb has a magnetic lever in an air gap that is connected to some external contacts. When we energize the left-hand coil, the relay contacts close, so energizing the centre-limb coil. This generates sufficient flux to keep the contacts closed regardless of the current in

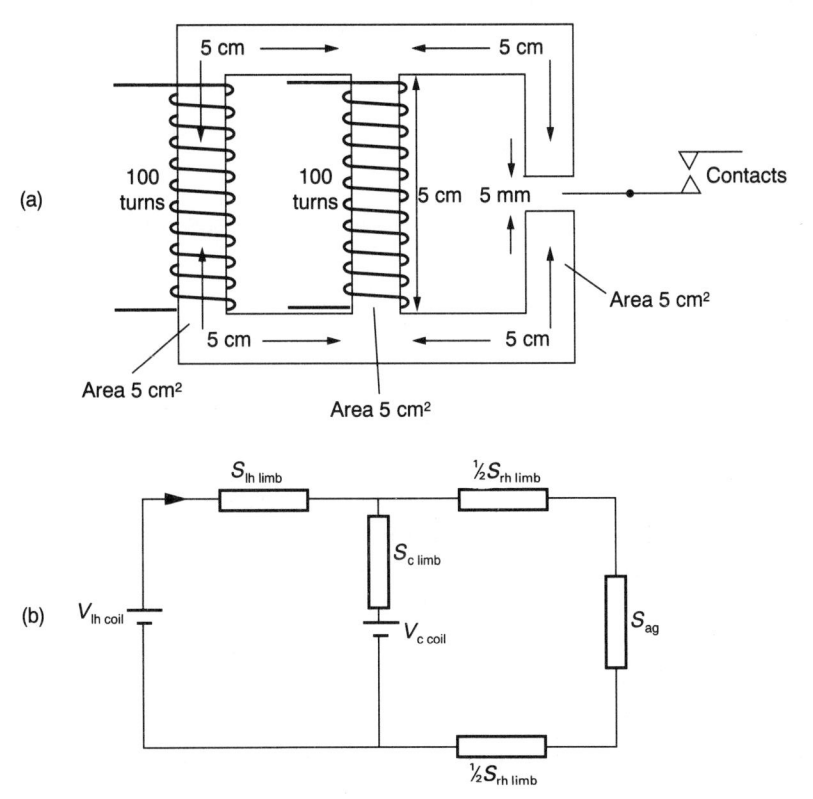

Fig. 7.7 (a) A two-coil relay actuator; and (b) electrical analogy for the two-coil relay actuator

the left-hand limb. We need a force of 0.5 newton to close the contacts, and the contact magnet produces a flux of 10 μWb. The relative permeability of the core is constant, and has a value of 100. We require to find the current needed in either coil to cause the contacts to close.

To analyse this situation, we will use an electrical analogy. Each coil is a source of magnetic potential and so we can represent them by voltage sources. We can represent flux by current, and core reluctance by resistance. Thus we can generate the electrical analogy shown in Fig. 7.7(b).

Now, we do not know the flux needed to close the contacts, and so this must be our starting point. From Section 3.11 we know that the force between two magnetic surfaces is (Equation (3.69))

$$\text{force} = \frac{1}{2} \frac{B^2}{\mu_0} \times \text{area}$$

where B is the flux density in the air gap. We require a force of 0.5 newton, and so we have

$$0.5 = \frac{1}{2} \frac{B^2}{4\pi \times 10^{-7}} 5 \times 10^{-4}$$

giving

$$B^2 = 2.5 \times 10^{-3}$$

So,

$$B = 5 \times 10^{-2} \text{ Wb m}^{-2}$$

Now, the area of the face generating this flux density is 5 cm^2, and so the total flux in the air gap is

$$\phi = 25 \times 10^{-6} \text{ Wb}$$

As the flux from the magnet is 10 μWb, the flux needed from the magnetic circuit is 15 μWb. So, the flux in the right-hand limb is 15 μWb.

We now need to determine the reluctance of the right-hand limb. As Fig. 7.7(b) shows, there are three components to the total reluctance: the upper half of the limb; the air gap; and the lower half of the limb. The length of this limb is 10 cm, in which there is a 5 mm air gap. As the air gap is very small compared with the length of the limb, we can write

$$S_{\text{rh limb}} = S_{\text{limb}} + S_{\text{air}}$$

$$= \frac{10 \times 10^{-2}}{100\mu_0 \times 5 \times 10^{-4}} + \frac{5 \times 10^{-3}}{\mu_0 \times 5 \times 10^{-4}}$$

$$= 1.6 \times 10^6 + 8 \times 10^6$$

$$= 9.6 \times 10^6 \text{ At Wb}^{-1}$$

7.6 The magnetic circuit

Thus the magnetic potential across the right-hand limb is

$$V_{\text{rh limb}} = S_{\text{rh limb}} \times \phi_{\text{rh limb}}$$
$$= 9.6 \times 10^6 \times 15 \times 10^{-6}$$
$$= 143 \text{ At}$$

This must be the potential across the centre limb, which is made up of the potential generated by the centre coil, and the reluctance of the centre limb. If we assume that the centre coil is not energized, this circuit only consists of the reluctance given by

$$S_{\text{c limb}} = \frac{5 \times 10^{-2}}{100\mu_0 \times 5 \times 10^{-4}}$$
$$= 8 \times 10^5 \text{ At Wb}^{-1}$$

Thus the flux down the centre limb is

$$\phi_{\text{c limb}} = \frac{V_{\text{c limb}}}{S_{\text{c limb}}}$$
$$= \frac{143}{8 \times 10^5}$$
$$= 1.8 \times 10^{-4} \text{ Wb}$$

This must be added to the flux down the right-hand limb and so the flux from the left-hand limb is

$$\phi_{\text{lh limb}} = \phi_{\text{c limb}} + \phi_{\text{rh limb}}$$
$$= 1.8 \times 10^{-4} + 15 \times 10^{-6}$$
$$= 1.95 \times 10^{-4} \text{ Wb}$$

This flux has to flow through the reluctance of the left-hand limb, which is

$$S_{\text{lh limb}} = \frac{10 \times 10^{-2}}{100\mu_0 \times 5 \times 10^{-4}}$$
$$= 1.6 \times 10^6 \text{ At Wb}^{-1}$$

Hence the magnetic potential is

$$V_{\text{lh limb}} = \phi_{\text{lh limb}} \times S_{\text{lh limb}}$$
$$= 1.95 \times 10^{-4} \times 1.6 \times 10^6$$
$$= 312 \text{ At}$$

This must be added to the potential across the centre/right-hand limb. So, the potential generated by the left-hand coil is

$$V_{\text{lh coil}} = V_{\text{lh limb}} + V_{\text{rh limb}}$$
$$= 312 + 143$$
$$= 455 \text{ At}$$

As there are 100 turns on the left-hand coil, the current has to be

$$I_{\text{lh coil}} = \frac{455}{100}$$
$$= 4.55 \text{ A}$$

Now, if no current flows through the coil on the left-hand limb, the centre limb has to supply the flux. As we have already seen, the flux down the right-hand limb is 15 μWb, and this must be added to the flux down the left-hand limb given by

$$\phi_{\text{lh limb}} = \frac{V_{\text{lh limb}}}{S_{\text{lh limb}}}$$

$$= \frac{143}{1.6 \times 10^6}$$

$$= 8.94 \times 10^{-5} \text{ Wb}$$

Thus the total centre-limb flux is

$$\phi_{\text{c limb}} = \phi_{\text{lh limb}} + \phi_{\text{rh limb}}$$
$$= 8.94 \times 10^{-5} + 15 \times 10^{-6}$$
$$= 1 \times 10^{-4} \text{ Wb}$$

and so the magnetic potential across the centre limb must be

$$V_{\text{c limb}} = \phi_{\text{c limb}} \times S_{\text{c limb}}$$
$$= 1 \times 10^{-4} \times 8 \times 10^5$$
$$= 80 \text{ At}$$

Therefore the coil on the centre limb has to supply

$$V_{\text{c coil}} = V_{\text{c limb}} + V_{\text{lh limb}}$$
$$= 80 + 143$$
$$= 223 \text{ At}$$

As the centre limb has 100 turns wound on it, the centre limb current is

$$I_{\text{c limb}} = \frac{223}{100}$$

$$= 2.2 \text{ A}$$

This example has shown how to analyse a magnetic circuit by using an electrical analogy. We were able to use the reluctance of the core because we assumed that the core had a linear B/H curve. If we cannot assume this, we would have to calculate B, find H from the curve, multiply by the length of the circuit, find the new flux, and so on. This is the method that must be used in problem 7.3.

PROBLEMS

Note: It is **essential** that you draw lots of diagrams to help you visualize the situation. Although some of the problems may seem trivial, they do have a purpose – to get you thinking in three dimensions.

CHAPTER 1

1.1 Determine the circumference of a circle, of radius r, by integration. (Use the cylindrical coordinate set, but ignore the z-axis. Take a small section of the circumference of length dl, subtending an angle of $d\phi$ at the centre of the circle. It is then a matter of integrating with respect to ϕ between the limits 0 and 2π.)   Answer: $2\pi r$

1.2 Determine the area of a circle, radius r, by integration. (Consider a small incremental area, ds. Take the thickness of this area to be dr, and let it subtend an angle of $d\phi$ to the centre of the disc. It is then a matter of integrating with respect to ϕ, and then with respect to radius.)   Answer: πr^2

1.3 Determine the surface area of a sphere, of radius r, by integration. (Take a small incremental area, ds, on the surface. Use the spherical coordinate set, and integrate with respect to ϕ and θ.)   Answer: $4\pi r^2$

1.4 Determine the volume of a cylinder of radius r and height l. (Use the result of problem 1.2, and calculate the volume of a disc of height dz. Then integrate with respect to z.)   Answer: $\pi r^2 l$

1.5 Determine the volume of a sphere of radius r. (Problem 1.3 gave the surface area of a sphere. It is then a matter of calculating the volume of an incremental sphere of thickness dr, and integrating with respect to radius.)   Answer: $\tfrac{4}{3}\pi r^3$

CHAPTER 2

2.1 Determine the flux radiating from a positive point charge of magnitude 400 pC. What is the flux density and electric field strength at a distance 10 mm from the charge? What is the force on a 10 µC point charge at this radius?
Answer: 400 pC; 0.32 r µC m^{-2}; 36 r kV m^{-1}; 0.36 N

2.2 A linear electric field of strength 10 V m^{-1} is established in free space. Determine the work done in moving a 10 µC charge a distance of 1 metre against the field. Repeat if the field makes an angle of 30° to the direction of travel. Answer: 0.1 mJ; 86.6 µJ

2.3 Determine the absolute potential at distances of 50 cm and 100 cm from a negative point charge of 20 µC. Hence find the potential difference between the outermost and innermost points. Answer: −360 kV; −180 kV; 180 kV

2.4 Two charges, each of magnitude 10 pC, are situated on the x-axis at 1 m and 3 m from the origin. Determine the flux density, the electric field strength, and the absolute potential at a point mid-way between them, and 1 m along the z-axis. (Use the principle of superposition, i.e. consider each charge separately. When you are dealing with vectors, resolve them into horizontal and vertical components, and then add components to find the resultant.) Answer: 0.6 z pC m^{-2}; 63.6 z mV m^{-1}; 127 mV

2.5 Determine the flux density, electric field strength, and absolute potential produced at a distance of 1 m from the centre of a 1 m radius disc with a charge of 10 µC evenly spread over it. Assume that the disc is in air.
Answer: 0.47 z µC m^{-2}; 52.7 kV m^{-1}; 74.5 kV

2.6 A length of coaxial cable is to have a maximum capacitance of 100 pF m^{-1}. The cable is to be air-cored with a maximum electric field strength of 3 MV m^{-1}. As the cable is to be wound on a drum, the maximum outer radius is restricted to 3 cm. Determine the radius of the inner conductor, and find the maximum voltage that can be carried.
Answer: 1.7 cm; 28.4 kV

2.7 A 200 m length of feeder consists of two 2 mm radius conductors separated by a distance of 20 cm in air. A potential of 1 kV is maintained between the two conductors. Determine the minimum and maximum values of field strength. If air breaks down at 3 MV m^{-1}, determine the maximum voltage that the feeder can withstand.
Answer: 55 kV m^{-1}; 2.2 kV m^{-1}; 55 kV

2.8 A 47 µF capacitor is charged to a potential of 50 volt. The capacitor is then suddenly connected to a discharged 10 µF capacitor. Determine the energy stored in the 47 µF capacitor before and after connection, and the energy stored in the 10 µF capacitor. What is the difference in total stored energy? (Remember that charge must be conserved, and that the capacitors are in parallel. The energy difference is due to heat loss generated by current passing through any lead resistance as the potentials are equalized, and in the spark that may occur when the capacitors are connected.)
Answer: 58.8 mJ; 39.5 mJ; 8.4 mJ; 10.9 mJ

2.9 A variable capacitor consists of a number of vanes that are free to rotate. A voltage of 10 volt is applied to the capacitor which has an initial value of 3 µF. If the capacitance is increased to 10 µF, determine the new potential. Compare the stored energy before and after the change. Answer: 3 V; 150 µJ; 45 µJ

CHAPTER 3

3.1 Determine the flux radiating from an isolated north pole of strength 400 pWb. What is the flux density and magnetic field strength at a distance 10 cm from the pole? What is the force on a 10 µWb north pole at this radius?
Answer: 400 pWb; 3.2 r nWb m^{-1}; 2.5 r mN Wb^{-1}; 25 r nN

3.2 A current element of length 1 μm passes a current of 1 A. This element is placed at the origin of a Cartesian coordinate set, and is lying along the vertical z-axis. Determine the magnetic field strength at a point whose coordinates are $x = 1$ m, and $z = 0.5$ m. Hence find the force on a 10 μWb north pole placed at this point

Answer: 57 ϕ nN Wb^{-1}; 0.57 ϕ pN

3.3 Find the flux density for the situation described in problem 3.2. If the magnetic monopole is replaced by a current element of length 1 μm and current 3 A, determine the force on it. (Note the exceedingly small force on the current element. This is due to the **B** field being used to find the force on a current element.)

Answer: 7.2×10^{-14} ϕ Wb; 2.2×10^{-19} ϕ N

3.4 An ideal capacitor, of value 100 pF, is connected to an a.c. source of 10 volt peak and frequency 200 Hz. Determine the current taken by the capacitor, and hence the magnetic field strength at a distance of 5 cm from the capacitor leads. If you apply Ampère's law between the plates of the capacitor, is there still a field? If there is, where is the current enclosed by the path? (Use Ampère's law here. As regards the magnetic field between the plates, it is there, and it is due to displacement current.)

Answer: 1.3 μA; 4.1 μWb

3.5 An air-spaced twin feeder transmission line carries a current of 1 kA. The conductors are spaced 5 metre apart. Determine the force per unit length between the conductors.

Answer: 0.04 N m^{-1}

3.6 Determine the inductance per unit length for the coaxial cable designed in problem 2.6. Hence find the self-resonant frequency of a 200 m length.

Answer: 164 nH m^{-1}; 197 kHz

3.7 A length of 4 mm wide microstrip is etched on one side of some double-sided pcb. The thickness of the board is 3 mm, and the non-ferrous dielectric has a relative permittivity of 3. Estimate the capacitance and inductance per cm, and the resonant frequency of a 5 cm length. Answer: 0.35 pF cm^{-1}; 9.9 nH cm^{-1}; 537 MHz

3.8 An alternative method to finding the internal inductance of a piece of wire is one based on energy storage. This equates the stored energy in the wire, obtained from a 'fields' point of view, to that obtained from a 'circuits' point of view. Use this method to obtain the internal inductance of a piece of wire. (Refer to Fig. 3.14(a) and find the magnetic field at radius r – see Example 3.2 in Section 3.6. Then find the energy per m^3 from Equation (3.67) and hence find the fractional stored energy in the tube of Fig. 3.14(a). It is then a case of integrating with respect to radius to obtain the total energy stored in the wire. This can be equated to that obtained from Equation (3.66) to get the inductance.)

CHAPTER 4

4.1 The mobility of an electron in a metal is given by

$$\mu = \frac{q\tau}{2m}$$

where τ is the mean time between collisions, and m is the mass of an electron

(9.11×10^{-31} kg). Determine the time between collisions for an electron in copper with 8.5×10^{28} free electrons per metre^3, and $\sigma = 58$ MS m^{-1}. Answer: 4.9×10^{-14} s

4.2 If the sample in problem 4.1 has a current density of 1×10^6 A m^{-2}, determine the electric field strength. Hence estimate the drift velocity of the electrons.
Answer: 17.2 mV m^{-1}; 73.5 μm s^{-1}

4.3 A 1 metre long cylindrical block of metal has a resistance of 5 μΩ and radius 5 cm. The block is drawn through a die into wire of radius 1 mm. Determine the resistance per metre length of wire. Answer: 12.5 mΩ

4.4 A length of coaxial cable has an inner conductor of radius 2 mm, and an outer conductor of radius 2 cm. The conductivity of the dielectric is 5×10^{-4} S m^{-1}, and it has a relative permittivity of 4. The cable carries a voltage of 1 kV at a frequency of 1 MHz. Determine the resistance of a 100 m length, and hence find the loss tangent. What power is lost due to the shunt resistance? Answer: 7.33 Ω; 2.25; 136 kW

CHAPTER 6

6.1 Show that the absolute potential at a point P, distance R from an electric dipole, is given by

$$V = \frac{p \cos \theta}{4\pi \epsilon R^2}$$

where θ is the angle made by a line joining P to the dipole. (Draw the dipole pointing upwards, with the point P to the right-hand side. The angle θ is the internal angle. Also, use the principle of superposition.)

6.2 A 1 V m^{-1} electric field, travelling in air, is obliquely incident on a block of glass, $\epsilon_r = 6$. Determine the angle the transmitted field makes to the boundary if the field makes an angle of 45° to the glass block. Answer: 9.5°

CHAPTER 7

7.1 A hysteresis loop has an area of 100 cm^2 when drawn with 1 cm representing 25 At m^{-1} and 1 cm representing 0.2 Wb m^{-2}. Determine the hysteresis loss per unit volume if the sample is used in a transformer operating at 50 Hz. Answer: 25 kW m^{-3}

7.2 The bobbin of a moving-coil meter has $40\frac{1}{2}$ turns of wire wound on it, and is placed in a magnetic field which has a flux density of 20 mWb m^{-2}. The bobbin is connected to an indicating mechanism with a restoring spring that exerts a torque of 1 μN m per degree of deflection. The coil is 5 cm long, and has a diameter of 3 cm. Determine the current required to give a deflection of 15°. Answer: 12.35 mA

7.3 Repeat the example of Section 7.6 using the B/H data given below:

B mWb m^{-2}	30	400	430	80	110
H kAt m^{-1}	1.4	5.2	6.0	2.6	3.1

Answer: 8.6 A; 4.15 A

Index

Ampère
  André 2
  definition of 77
Ampère's law 75

Biot–Savart law 69
Boundary relationships
  electric field 158
  magnetic field 166
B/H curve 164

Capacitors 36–58
  coaxial cable 38
  dielectrics 150
  displacement current 50, 127
  electrolytic 162
  energy storage in 47
  force between plates 49
  leakage resistance 126
  loss tangent 128
  low frequency effects 50
  microstrip 45
  parallel combinations 56
  parallel plate 37
  resistance to electric flux 54
  series combinations 57
  twin feeder 41
  wire over ground 44
Cartesian coordinates *see* Coordinate systems
Cathode ray tube 58
Charge 1, 3, 8
  density 24
  line 23
  point 10
  surface 29
  volume 35
Coaxial cable
  capacitance 38
  inductance 98
  series resistance 130
  shunt resistance 130
Coercivity 164
Coils 79–94
  rf choke 110
  simple 93
  solenoid 83
  toroidal 88
Conductivity 123
Conductors 3, 120
  conductivity 123
  drift velocity 123
  mobility 123
Coordinate systems
  Cartesian 5
  cylindrical 5
  spherical 5
Corkscrew rule 67
Corona discharge 161
Coulomb, Charles Augustin de 1
Coulomb's Law 8, 16, 64
CRT *see* Cathode ray tube
Current 3
  definition 121
  density 123
Cylindrical coordinates *see* Coordinate systems

Dielectric strength 160

Dielectrics 150–162
  break down 160
  dipole moments 150
  electric dipoles 150
  flux density 154
  polarization vector 153
  saturation 150
  susceptibility 154
Dipole moments
  electric 150
  magnetic 163
Displacement current 50, 127
Drift velocity 123

Electric dipoles 150
Electric field 13
Electric field strength 14, 122
  boundary relationships 158
  maximum 160
Electric flux 10
  boundary relationships 158
  density 12
  distribution 11
Electric potential 16, 121
Electroconduction 120–142
  capacitors 126
  coaxial cable 130
  conductivity 123
  current density 123
  current flow 120
  drift velocity 123
  microstrip 133
  mobility 123
  Ohm's law 125
  power loss 125
  resistance 125
  twin feeder 133
Electrolyte 162
Electrolytic capacitors 162
Electromagnetism 67–119
  Ampère's law 75
  Biot–Savart law 69
  coercivity 164
  corkscrew rule 67
  current element 70
  deflection of an electron
    beam 111
  Faraday's law 92
  field due to a coil 79
  field strength 71
  flux 68
  flux density 71
  flux linkage 92
  force between two coils 107
  force on a conductor 70
  hysteresis loss 169
  inductance 92
  magnetic potential 90
  plasma confinement 114
  reluctance 89
  remnance 164
  solenoid 83
  toroidal coil 88
Electron 3
Electron–volt 60
Electrostatic precipitator 60
Electrostatics 1, 3, 8–63
  Coulomb's law 8
  energy density 48
  field strength 14
  flux 10
  flux density 12
  flux distribution 11
  force between charged plates 49, 147
  polarization vector 153
Energy density
  electrostatic 48, 146
  magnetic 106, 146
Energy storage
  electrostatic 47, 146
  magnetic 105, 146
Equipotentials
  lines 21
  surfaces 21

Faraday, Michael 2
Faraday's law 92

Ferromagnetic materials 163–176
  coercivity 164
  hysteresis loss 169
  magnetic dipoles 163
  magnetic polarization vector 164
  remnance 164
  saturation 164
Field strength 145
  electric 14
  magnetic 65
Flux
  electric 10, 144
  magnetic 65, 68, 144
Flux density
  electric 12, 145, 154
  magnetic 65, 145
Force
  between charged surfaces 49, 147
  between current-carrying wires 77
  between magnetic surfaces 107, 147
Force fields 143

Galvani, Luigi 1
Gaussian surface 12, 23
Gauss' Law 12, 36, 65
Generators 171

Heaviside, Oliver 2
Helmholtz, Hermann von 83
Helmholtz coils 81
Henry, Joseph 93
Hertz, Heinrich Rudolf 2
Hysteresis loss 169

Inductance 92–111
  back emf 92, 110
  coaxial cable 98
  energy storage 105
  isolated wire 94
  low frequency effects 108
  microstrip 103

rf chokes 110
self 94
simple coil 93
skin effect 99
twin feeder 100
Insulators 3
Integration 5
Inverse square law 9
Iron *see* Ferromagnetic materials

JET – Joint European Torus 114

Kirchhoff's current law 137
Kirchhoff's voltage law 136

Loss tangent 128

Magnetic dipoles 163
Magnetic field 64
  boundary relationships 166
  strength 65
Magnetic flux 65
  boundary relationships 167
  density 65
Magnetic materials *see* Ferromagnetic materials
Magnetic potential 90
Magnetism 2, 4, 64
  permanent 4
Magneto-motive force *see* Magnetic potential
Marconi, Guglielmo 2
Maximum power transfer 169
Maxwell, James Clerk 2
Metals *see* Conductors
Method of images 44
Microstrip
  capacitance 45
  inductance 103
  series resistance 133
Mobility 123
Monopoles 64
Motors
  torque generation 170

Oersted, Hans Christian   2, 67
Ohm, Georg Simon   125
Ohm's law   125

Permanent magnet   164
Permeability
  free-space   65, 66
  relative   66, 164
Permittivity
  free-space   9
  relative   9, 153
Polarization vector
  electric   153
  magnetic   164
Potential   146
  absolute   19
  difference   18
  electric   16, 121
  magnetic   90
Potential energy   16
Proton   3, 8

Reactance
  of a capacitor   51
  of an inductor   110
Relative permeability *see*
  permeability
Relative permittivity *see* permittivity
Reluctance   89
Remnance   164
Resistance to flux   146
Resistors   124–139
  capacitors   126
  coaxial cable   130
  integrated circuit   139
  microstrip   133
  parallel combination of   137
  series combination of   138
  twin feeder   133
Rf chokes   110

Saturation *see under* Dielectrics,
  Ferromagnetic materials
Scalar   4
Skin effect   99
Solenoid   83
Speed of light   79, 103
Spherical coordinates *see* Coordinate
  systems
Surface charges   158
Surface currents   166
Susceptibility   154

Toroidal coil   88
Transformers   167–170
  B/H curve   164
  hysteresis loss   169
  impedance matching   169
  iron-cored   163, 167
  magnetic circuit   173
  open- and short-circuit
    testing   169
  step-down   168
  step-up   168
Twin feeder
  capacitance   41
  inductance   100
  series resistance   133

Vectors   5
Volta, Count Alessandro   2, 18
Voltaic cell   2

Weber, Wilhelm Eduard   65